THE HERESY OF OEDIPUS AND THE MIND/MIND SPLIT

A Study of the Biocultural Origins of Civilization

Maria M. Colavito

The Edwin Mellen Press
Lewiston/Queenston/Lampeter

Library of Congress Cataloging-in-Publication Data

This book has been registered with the Library of Congress.

ISBN 0-7734-8854-5

A CIP catalog record for this book is available from the British Library.

The Edwin Mellen Press
Box 450
Lewiston, New York
USA 14092-0450

The Edwin Mellen Press
Box 67
Queenston, Ontario
CANADA L0S 1L0

The Edwin Mellen Press, Ltd.
Lampeter, Dyfed, Wales
UNITED KINGDOM SA48 7DY

Printed in the United States of America

DEDICATION

In loving memory of my father,
Who taught me to love the myths.

And for my mother,
Who forced me to live them . . .

TABLE OF CONTENTS

ACKNOWLEDGMENTS

My acknowledgments run in reverse chronological order, since it seems to be the trend of this book . . .

Thank you to my students at Flagler College, especially the "philosopher chefs," who inspired me to finish this work because they wanted to know. You should be advised that the ban on "writing in the snow" is officially lifted, with the publication of this book.

I also offer my gratitude to my undergraduate professors at the State University of New York at Stony Brook, who showed me what true, original scholarship was; especially Antonio T. de Nicolas, Joan Fry, Michael Gazzaniga, and Richmond Hathorn, none of whom are still there. At least I managed to keep their company with me through the writing of this book; actually, some closer than others . . .

And first chronologically, I'd like to say thanks to my family; to my sister Adrienne and brother Carl, who had always managed to offer some cheering on when I just couldn't seem to get it together; to my mom, who made me learn to express my creativity and gave me the discipline to always finish what I start, including this book; but most of all, to my dad, who urged me "onward to the Peak!" Well dad, how am I doing?

The Heresy of Oedipus
And the Mind/Mind Split:

A Study Of The Biocultural Origins of Civilization

PREFACE

This book, *THE HERESY OF OEDIPUS AND THE MIND/MIND SPLIT*, by Dr. Maria Colavito, needs an introduction the way Darwin's *THE ORIGIN OF SPECIES* does. The reader should be warned that what he/she is about to engage in is not a systematic accumulation of facts, though the book is filled with them; nor another grandiose theory of how things would hang better together, though there are many of those theories buried for ever by this book; nor even a unified field theory of the historical human species, though the human species comes out at last bioculturally unified.

What this book brings to us is a new model, a paradigm on how the human species has bioculturally acted on itself and the environment through all its historical existence. Where Darwin stops with the evolution of species, Dr. Colavito begins with biocultures as the biological evolution of the human brains and their historical development on the planet. Biocultures, in this context, is the same as biological foundations (the historical evolution of the human biological brains) giving rise to their own cultural extension in human individuals and communities. According to Dr. Colavito, the invariant brains of humans: the reptilian, limbic, the right and the left hemispheres of the neocortex, and finally the "interpreter module," are the biological invariants of the human species. It is a biological fact that these brains did not appear simultaneously in humans but evolved according to need or exercise, as they do in every new-born infant, building themselves as neural paths in the brains and as external realities or cultures for the humans who used them as they became available. Thus, we know of ancient cultures as being (using Dr. Colavito's terminology) maia types (kinesthetic), since the brain at their disposal was primarily the reptilian; or mythos types (auditory-affective) since they already had developed the limbic brain; or right brain mimetic since they acted on the language of images of the right hemisphere of the neocortex (magicians, leaders, the demiurge); or left brain mimetic (theoreticians, theologians, social scientists) since they acted primarily from the left hemisphere of the neocortex; or simply logos (the last to evolve) since these types are imageless, caught in the loop of verbal skills unable to transcend to the right brain even when they depend for all their information on that side of the brain.

Following Dr. Colavito's laid-out development, these five biological invariances give rise to five epistemological invariances, the cultural invariances as foundation for the types described as maia, mythos right brain mimesis, left brain mimesis, and logos; that is, individual humans or human communities where one of those five types is the predominant bioculture of the individual or the group. These types are invariant and they have a history.

The "history" of the five biocultures is the scope of this book. The challenge to the reader is to be able to follow the development of the model without jumping the gun with irritation or misreading (your bioculture will be showing) but by reading the book to the end, finding the reader's own bioculture and reading the book again looking for the ways of others to find oneself and within the human communities around us without ideology but with heart.

Biology will find little quarrel with this new model. Contemporary studies on the biological development of the brain of infants and split-brain research prove it daily. Now that the model is manifest as a model, advances in neurobiology will probably become routine. In view of these studies, the education and nurture of infants and young children will become paramount since the lack of development of biocultures in children up to the age of twelve, in some cases and much earlier in most, is irreversible. In other words, if the child's reptilian and limbic brains are not exercised early, as early as the first months of their life, they do not develop neural chains for the life of these brains and atrophy follows. The "windows of malleability" are closed forever. By age twelve, all the brains are either present or some are absent and this process is irreversible. In most cases, the child ends up left brain mimetic, with luck, or simply logos, a victim of the loops of his or her own verbal skills as in a jail, unable to reach either the right brain world of images or the outside world. I remember a student of mine who approached me after class to question what she called my interpretation of the people of the Rig Veda. According to her, it was not possible for people to see images with the brain much less make them. I tried to explain to her that images are something everyone makes or has. For instance, I asked her, what is the image of your own mother when you are not with her. Her answer left me cold. "The only image of my mother I carry with me is m-o-t-h-e-r." Spelling is as far as her brain could imagine. This example, I found subsequently repeated by many other young people.

The biocultural model of humans is not, of course, what we have been told rules the human species. The model of the Ascent of Man, the climb of the faculties up the ladder of movement, feeling, auditory, image, REASON, the rule of reason over the chaotic field of the senses, the rational over the emotional, this is more likely to sound familiar to us. Let one brain be the rule over the many. How did this story become history, the history of the human species, one brain, one God, one culture, one global unity, one history?

THE HERESY OF OEDIPUS develops in detail this perverse reading of human evolution. The story begins before the Flood when the Phoenicians ruled the waterways and by inventing writing, they introduced in the species a new technology that of necessity changed the neural paths of the human brains. While pictographic writing accounts for the mind/body split and the left brain development, alphabetic writing accounts for the "interpreter module" and the mind/mind split. The rest is history. This history, however, is the history of the Phoenician people (the mind/mind split) and written by Phoenicians already with the mind/mind split. Their reading could not be otherwise. Oedipus is the mind/mind split hero and with him the whole house of Cadmus is destroyed. The Oedipus of this book is not Freud's Oedipus, nor is he the hero of the individual freedom. He is just the prototype of the logos people and of the devastation they bring with them for the rest of us. All our "windows of malleability" are in mortal danger of being shut closed.

This extraordinary and compelling model of the human species as described and verified by Dr. Colavito in this book would not have been possible as a model had she not made another crucial discovery. That next to and based on the biological invariants of the brains there are five invariant epistemologies on which all philosophy and description, scientific or otherwise, is based. This is a novel discovery in an age when philosophy is victim to the demand that epistemology be just a theory of knowledge. No more. Basing her discovery on some work of mine on imagination as found in earlier cultures like the Rig Veda, Plato and some mystics, Dr. Colavito has been able to enlarge my project to Greek, Phoenician and modern logos cultures to describe the invariant epistemologies on which they rest. As it turns out, the model takes the brains as a "primary text" upon which a "secondary text" is imprinted through repetition and exercise creating the neural passages (a biology). A "primary technology" is then applied to activate and set in motion the different biological brains while a "secondary technology" is used for reading the previous technology and texts. If the reading does not coincide with the technology in use, we have ideology and cultural imperialism, which is the actual history of history, or the Phoenicians (those of the mind/mind split), writing themselves and demanding that others do AS THEY DO. The outside world, what we call reality, is an extension of the brain in use, or the brain in use extends itself to form worlds, holograms that become the world, or reality.

Dr. Colavito is relentless in the pursuit of her model, like Yeats's falcon "turning and turning in the widening gyre, the falcon cannot hear the falconer," the ideological idol chases ideological idol to the dust and hardly any belief is left standing; history, philosophy, religion, medicine, the social sciences all shake in their foundations and are about to tumble and one would anticipate fall and leave us all bereaved with nothing to hold on to. Yet, magically, by the end of the book, nothing has been destroyed, the world continues to breathe and laugh and the sciences, the human institutions, the gods are again alive though none of them occupies the same blind space as before. There is a transformation and despite the pain, there is radiant light at the end of the tunnel. The reader, however, must be warned to place himself or herself in the space where this book is written; one space before information, one space before the several disciplines it describes, one space before any names of humans or gods. The place to read this book from is exactly the origin itself of human expression, the epistemological invariant through which the multiple human brains are and have been the instrumentation of human communication. Reading the conditioning of biology on the human species in no way claims "determinism" of the human will. Just the opposite, as our individual irreconcilable biographies attest to our human freedom.

Dr. Colavito is due a debt of gratitude by all of us. Our human world will not be covered by as many shadows and pains as before she entered the academic world. Her previous books, *THE PHYTHAGOREAN INTER-TEXT IN OVID'S METAMORPHOSES* (Mellen Press) and THE NEW THEOGONY *(*SUNY Press) established her as an authority in myth and the mythic imagination. This book, however, sets her apart on a space by herself. Here, she has spoken for the species at large or the species has spoken through her through a model that has always been present, had it not been for the interference of Oedipus and the forced ideological reading he brought on all of us with the mind/mind split and the Phoenician penchant for ideology.

Dr. Colavito is a graduate of the State University of New York at Stony Brook where she studied with Michael Gazzaniga in psychological studies, Richmond Hathorn in mythology, and myself in philosophy, epistemology, and Oriental studies. She is, no doubt, the best Stony Brook has produced at a time when some of the best were on its faculty. Those of us who worked with her are grateful, as I am sure the readers of this book will be.

❁ ❁ ❁

Antonio T. de Nicolas
Professor Emeritus of Philosophy
State University of New York
at Stony Brook

The Heresy of Oedipus And the Mind/Mind Split:

A Study Of The Biocultural Origins of Civilization

INTRODUCTION

The nature/nurture controversy, sometimes known as the evolution/environment controversy, seems to have trickled down into the information systems of the vernacular world as an unfortunate rift between dueling scholarly camps. In one camp, the nature/evolution camp, lie those scholars whose research on the forefront of neurobiological discoveries have led to amazing revelations about the biological origins of many of our "human" characteristics.[1] In the other camp, the nurture/environment camp, lie those scholars whose research on the forefront of sociological discoveries have led to amazing revelations about the sociological origins of many of our "human" characteristics. The glaring negative implications of wholeheartedly ingesting the evolutionary model seem obvious. Where do these evolutionary "discoveries" place our time-honored beliefs in human freedom, will, choice, individuality and the like? What of "original sin," moral judgement, reward and punishment, not to mention the "work and you too shall succeed" tenet? Finally, does accepting this evolutionary model, at root, mean accepting that we, as humans, are basically composed merely of deterministic, unconscious quanta? While few of us would truly espouse these coldly deterministic conclusions, most would also aver that accepting the basic tenets of the proponents of the environmental paradigm wholeheartedly likewise smacks at the face of reason, especially in a society that prides itself on state-of-the-art scientific discoveries as its primary criterion of superiority.

The biocultural paradigm is offered as a model that transcends both camps, by recognizing the neurobiological origins of human development AND by delineating exactly how sociological influences and when sociological influences can and cannot affect those neurobiological invariants.

Utilizing existing discoveries in evolutionary neuro-biology and Selection Theory, the scientific basis of the Biocultural Paradigm is established. This Paradigm is composed of five proto-cultural models ("biocultures") which correspond to the five evolutionary centers of our neurological structures; our five "brains." They are:

—*The Maia Bioculture*, whose primary neural link is the Reptilian brain[2] and its communication to the amygdala and septal regions of the Limbic System.

—*The Mythos Bioculture*, whose primary neural link is the prefrontal cortex and the Limbic System[3].

—*The Right Brain Mimetic Bioculture*, whose primary neural link is the right hemisphere[4] of the Neo-cortex.

—*The Left Brain Mimetic Bioculture*, whose primary neural link is the left hemisphere[5] of the Neo-cortex.

—*The Logos Bioculture*, whose primary neural link is to the "interpreter module" of the left hemisphere of the Neo-cortex[6].

Each bioculture, then, is formed by the cultural manifestation of the primacy of one neural function over the others. While, initially, it is through individual human development that choices are made as to the primacy of certain neurological traits over others; eventually, (through the habitual repetition of the primacy of certain neurological links at the expense of others), these individual traits become societal ones; thus forming the Socio-biological basis for the Biocultural Paradigm.

The sociological evidence for the Biocultural Paradigm is founded primarily upon Socio-linguistic grounds, by analyzing the relationships between the literary remnants of certain cultures and their corresponding social, political and religious structures. The literary evidence is examined as it elucidates a neural map of cortical activity (thereby offering clues as to the biocultural slant of the group), while the social, political and religious systems are examined for evidence of neurological predispositions that manifest externally as cultural substitution systems. And here, is where our thesis in this book originates. What if WHAT WE KNOW IS MERELY A DESCRIPTION OF HOW WE KNOW, BIOLOGICALLY? What if WHAT TO US POSES AS CULTURAL HISTORY IS REALLY SIMPLY THE ARCHAEOLOGICAL ARTIFACTS OF OUR EMERGING NEUROBIOLOGICAL SELVES? And finally, WHAT IF WE COULD TRACE THIS HUMAN HISTORY OF BIOLOGY THROUGH OUR OWN LITERARY REMNANTS OF THE PAST? Well, it seems that we can, and the biocultural paradigm does just that.

This book begins with an analysis of our TWO human language systems: the language of images and the language of concepts. Likewise, we humans are able to know in both of these ways: either mythopoetically or ideographically. The former language system relies on the primacy of the right brain, seat of the mythopoetic mindset, which functions by making images through the acts of imagining and fantasy. It also directs and synthesizes the sensory data from the other brains. Following this path of the right brain, most mythologies too recount the pattern of this brain activity and linguistically recall the functions of locomotion, emotion, acts of will, fights against fantasy and the quest for immortality.

Unlike the right hemisphere, the left ideographic brain, father of the left brain mimetic and logos biocultural types, can operate independently and therefore can be disembodied from its corresponding brains. It reacts to sensory data as filtered through the sieve of the right hemisphere and it perceives linearly and sequentially, translating the images and emotions and involuntary reactions from the other brains into conceptual frameworks and linguistic representational systems about those concepts. For purposes of this book, the mind/mind split then is the phenomenon whereby we, as humans allowed, through the voluntary repetition of certain human technologies, the left hemisphere of the neo-cortex to gain not only supremacy but actual autonomy over the other brains and correspondingly, over the body.

This split[7], the mind/mind split happened late in our human development. Our evolutionary history reveals that our left hemisphere is the more recent, but is it necessarily the most advanced for our species? The right/left brain schism has its own mythology: the myth of the fall, the fall from consciousness to conscious reflection, to a life of sequential time, to the death of images, and with that, the death of the human image as well. When the human retains the capacity for joint autonomy within the neo-cortex (when both hemispheres are able to be exercised in a culture), the ensuing result is always the inevitable mind/body split, for there does not have to be a conscious connection between the left side of the brain and the body. This crisis, the mind/body split has become a philosophical appropriation of this disembodied vision of the left brain. And its creation is very clearly elucidated in myth.

Like the mythopoetic cultures, the ideographic, left brain cultures also create their universe by the criteria of the mind functions through which they operate. The trouble is, there are no images in the left brain, no gods, no genuine feeling, no past, no creation. One might venture to say, nothing that we may truly call human. And we, as a species are on our way to universalizing this mode of the mind to the virtual exclusion of the others. And with it, our species' descent.

In Greek myth[8], at the close of the Bronze Age, the gods create a race of heroes to carry the culture forward again so that, even though the Iron Age will come and destroy most humans, there is still hope for others, thanks to the group of humans who plant the seeds of regeneration by providing us with all the models for human action. Who are our heroes? What is their training? Is it possible for our human model, the computer, to plant our seeds of regeneration? Would we ever want it to?

The origin of the biocultural paradigm lies in our human understanding of the role of language[9] and the methods by which our perceptions about language literally shape our culture, self-image and even universe. In the first chapter, the language of images, as a separate and distinct form of communication from that of conceptual language is introduced and discussed in relation to our neurobiological inheritance as a species as well as our corresponding cultural propensities. Here, the five major biocultural types are outlined and discussed. The link is established between the biological origin of certain human functions, including the neurological and neurolinguistic aspects, and the resulting cultural consequences of these functions. This chapter concludes with an analysis and explanation of how comparative biocultural types might respond to a series of culturally based issues such as: views of the afterlife, forms of religion, educators of the culture, and the like.

The second chapter begins with an analysis of Creation Myths from varying cultures to uncover first, that all creation myths inherently carry within them at least the mythopoetic bioculture origins: maia, mythos and right brain mimetic. Remembering that we distinguish culture types here by the criteria of their neurobiological origin, these myths are primarily a recounting of this neurobiological development in an effort to link our own acts of creation with the creation of the universe at large. While it is true that all five culture types are present in myths of creation, how these five are articulated forms the crux of the overall cultural slant of the group. Those myths that emphasize the language of images create a holographic view of reality such that human action and re-creation of the creative model yields predictable worlds which can be communally shared through those linking images. A culture thus formed bases its affinity upon a shared experience and a uniform language based upon shared images. Those cultures that are formed by conceptually conceived myths follow a different path. A consensus must first be reached among members as to the correct conceptual framework. Once the nameless group has systematically articulated this framework, or theory, the prescriptive and proscriptive behaviors, as well as the political, economic, religious, and social aspects of the group become re-interpreted such that ONLY within the parameters of this framework is any human experience validated. Marginal society is formed by those who either do not wish to or are incapable of BIOLOGICALLY adapting to the new infrastructure. Hence the birth of biocultural imperialism.

Chapter Three focuses on the myth of Oedipus, perhaps the most blatant vestige of the result of the two biocultural types (the mythopoetic and the ideographic) co-existing in a single city. The blind seer Tireseas represents the mythopoetic world; Oedipus is the ideographic usurper. Together their paths weave an inextricable pattern of intrigue such that even we, as modern readers are not sure just exactly how to take the story's outcome. Somehow, though, we always seem to side with our Oedipus. So what is it about this character that has us all reeling? The Oedipus of Sophocles is perhaps the most famous figure in history because of his heroic insistence on following his own logical premises through to their conclusion, albeit at the expense of destroying city, home, family and gods, not to mention his own sight; in fact the very vehicle that propelled him into his logical dilemma in the first place. Yet, we, despite these negative ramifications, like Sophocles, applaud his audacity to snub his nose at the face of the gods and take his licks because they're HIS: not because of the gods, nor the fates, nor a providential encounter. He took his life in his own hands. It was more heroic of him to cause his own blindness than to succumb to the outmoded model of fate. And we see this as a victory of the self. And we see this as a model of the liberated human. And we see this as the ultimate act of defiance: free will. The trouble is that we see all these things in Oedipus BECAUSE WE SEE HOW HE SEES; that is, because we are viewing the myth with the "colored glasses" of our own logos bioculture. We love Oedipus because we have become Oedipus as a culture. But who was the real Oedipus? And why should he be considered the greatest heretic of the human family? The answer lies in the mind/mind split. Long before Rene Descartes inspirationally exclaimed, "I think, therefore I am", the human species was on the throws of a violent transformation. Long before modern philosophical/theological savants pondered the seemingly invariant human dilemma of the mind/body split with all of its cultural baggage, the species underwent a much more serious split in the shift from language dominance based on images to language dominance based on concepts; from the right hemisphere of the neocortex being the seat of the souls, to the left hemisphere being the seat of reason. Plato's articulation of the three souls gives us a small inkling into the nature of this earlier right brained dominant human. And the history of the house of Cadmus from where Oedipus is born, reveals how the birth of the logos bioculture gained a foothold in the world. So, why is Oedipus the heretic? Because he is the exemplar of the first logos bioculture type. He is the prototype of the human machine. He is the "reader of his own mind" to the exclusion of his ancestral history. And he is the first victim of it, too.

This fourth chapter elucidates how the exclusivity of the logos culture type, which pervades modern society, threatens to destroy much of our biological inheritance. The argument is simple: if our neocortex is dominated solely by left brain functions, as occurs when only abstract, logical modes of intelligence and learning are supported by the cultural group in power, the other neurological modes shall atrophy for lack of exercise. Evolution has shown us what happens to species whose development thus mutates. A short journey through history reveals how easy it is for cultural imperialism, armed with this biological weapon, to unwittingly destroy our human diversity, thereby destroying us as a species, for diversity is the very essence of our biological superiority.

❁ ❁ ❁

The charts and exercises found in the appendices are meant to be performed in conjunction with their corresponding chapters, in an effort to make visible, that which is merely intelligible through the simple reading of this book.

And finally, this book is not intended as an indictment against the biocultures of the left brain. It is rather a positive attempt to link all culture types by introducing them to our whole biological inheritance embodied in our many brains and manifest in our many cultures.

CHAPTER ONE

THE BIOCULTURAL PARADIGM

The story is that in the region of Naucrates in Egypt there dwelt one of the old gods of the country . . . Theuth. He it was that invented number and calculation . . . and above all writing. Now the king of that country at that time was Thamus, who dwelt in the great city of . . . Egyptian Thebes. To him came Theuth, and revealed his arts, saying they ought to be passed on to the Egyptians in general . . . . When it came to writing, Theuth said, "Here, O king, is a branch of learning that will make the people of Egypt wiser and improve their memories; my discovery provides a recipe for memory and wisdom." But the king answered and said, ". . . If men learn this, it will implant forgetfulness in their souls; they will cease to exercise memory because they rely on that which is written, calling things to remembrance no longer from within themselves, but by means of external marks . . . . And it is no true wisdom that you offer your disciples, but only its semblance, for by telling them of many things without teaching them you will make them seem to know much, while for the most part they know nothing, and as men filled, not with wisdom, but with the conceit of wisdom, they will be a burden to their fellows."

Plato, Phaedrus (274c-275b)

Once upon a time, I was teaching a Remedial Composition class at a local community college. In an effort to encourage my students to learn to think as well as write, one of the daily exercises in the class was a question of the day, where for twenty minutes the students would respond first in writing and then orally to a generic question posed. Later, I would collect the answers and grade the responses for grammar, syntax and diction; the customary things Composition professors are likely to do. This seemingly harmless exercise yielded for me unparalled revelations, not only about the nature of our "remedial" students' capacities, but moreover, about the nature of "composition" in general. In short, this is what happened: across the board, there was no comparison between the rich, structurally cohesive responses orally delivered by the students in class and the fragmented, seemingly ill-conceived written testimony of that response that was put on paper. The problem, it seemed, was certainly not in composition. So, where was it?

Well, the next day I brought up this issue in class. Why, I asked, is there such a difference between your written work and your oral presentations? This is how they explained it to me. The question "goes around in your head" and you feel like what you want to say. Then your body goes through the motions of how it would be if that answer happened. Then, if it feels like it's right to you, you say that. WHAT YOU PUT ON THE PAPER is whatever you need to TO REMIND YOUR BODY OF HOW YOU FEEL about the question. The only purpose of the actual words on the page, then, is to help you remember in your body the feeling of why you answered how you answered and no more. I, of course was astonished, but no more than they were when I told them my interpretation of the purpose of the words on the page, which at root, were essentially the same, with the added discomfort of the rules of grammar, syntax and diction. Simple enough for us, but to them, this response opened up a virtual Pandora's jar of questions. Why follow rules of grammar if they interfere with the flow of the feeling? Making the answer follows rules already, why add more rules? If the story comes out in one second, why do we have to explain IN ORDER all the events preceding the story if that's not how the story came to us? So, actually you want us to do two answers, one when we find the answer and another when we write it down? They just didn't get it, I thought. So, then we came to an agreement: I'd teach them the rules and they'd teach me their method of composing. Needless to say, within six months these students placed out of Remedial Composition and some even received A's in their regular English classes the next semester! And I, of course, was left to further ponder this issue.

So, what was the problem? What were their remedial skills? Simply put, translation. They could not translate their own images into the conceptual framework necessary to perform written skills. Once the problem was recognized, we as a class devised skills or short-cuts to the writing process that allowed them the ability to link the concepts in the words themselves to the images they were creating,(no small feat) and thereafter construct the written work. This, I might add, did not come naturally to them, though they were eventually able to learn the techniques. And I realized in the course of this experience that they were right; they did have to answer the question twice every time they wrote something. And in the process of working with them another element came into focus, one whose full import I only now am beginning to understand. Curiously enough, the majority of these students were also right brain dominant[1].

This experience rested in the recesses of my all too cluttered mind until one day, while reading Plato, I came across the above myth of Theuth. Suddenly, my mind was flooded with questions. What if it were true about writing, that it caused the opposite of what it portends to cause? How do we KNOW we know, if all we can do is ingest and recall, catalog and consume data as filtered through the written word, usually with little actual, first hand, eye witness knowledge? Do I really know what the world situation is like if I've never been to a country where people are starving, but I send a check because I read about it in the paper? Or maybe I sent that check because of the image of that skeletal phantom of a child with protruding abdomen and flies covering his little face . . . It was definitely the picture. But even this picture, driving me as it were to the depths of human compassion, was only a shallow vestige of the reality that was/is that child somewhere, yet it alone was sufficient to set my own body into a series of responses to hunger itself as if it were happening to me. And then I sent the check. There, but for the grace of God, go I . . . A trillion words, no action; a single picture, the world responds. What's going on here?

✿ ✿ ✿

The Two Languages of the Mind: Mythopoetic and Ideographic

It seems to me that what was going on might easily be described as the convergence of two distinct mind languages; I'll label them, for clarity, mythopoetic language and ideographic language. Let me, at the outset discourage any similarity between what is about to be said and the numerous works that have been produced on the issue of orality and literacy as it relates to language systems, per se[2]. Unlike those works, this study is not meant to be a discussion about the written manifestations of language; it is rather about the cerebral functions that translate themselves INTO language. And my contention is that there are two neural pathways to language. Mythopoetic[3] is a term which engenders those modes of knowing, apprehending, perceiving, or comprehending without the medium of concepts. In other words, mythopoetic language creates primarily in images; on the audial, visual, kinesthetic, olfactory, and gustatorial sensations that, membered and then dismembered form images in the minds of humans. Ideographic[4] language, then refers to the complex thought processes created as a result of a conclusion drawn from the already formed image. Here the sensory data has already been interpreted and the image formed. Indeed the image becomes a single quantum of information on which, or about which or through which some other piece of information shall be based. That other piece of information then, is the concept.

Here's a rudimentary example, using the medium of the written word, of the difference between mythopoetic and ideographic language:

IDEOGRAPHIC: The dog was hit by a car.

MYTHOPOETIC: A screeching bellow and a deep thump. Splattered and oozing red trickling over the matted hair on the mangled carcass.

Which way of receiving the information makes you wince? My guess is the mythopoetic, right? So, now the big question. Why? Because in order to receive the information, you had to make the dog, then make him get hit. In other words, you had to literally create the image, give it life and then, in this case, destroy it. Since you made it through your own sensory modes, you also felt its demise through those same modes. In other words, the dog is not some theoretical construct out there; it's in you. In the mythopoetic mode of language, you don't assume a dog, you make a dog. And it's a whole lot harder to live with the piece of information received thus than it would have been had it only been received ideographically. But then, if all information were merely recounted in the ideographic mode of language, imagine how boring our lives would be: constantly ingesting data, transferring and processing tidbits of facts here and there at the appropriate times in conversations, inventing "what if" circumstances and the following them through to their conclusions for the fun of it, writing laws, rewriting laws, exchanging vows, renouncing vows, ingesting new data, reading a newspaper.

> One headline: 300 Dead. Another Headline: Braves Win. Another Headline: Dog Hit By Car. Another Headline: Education on Decline . . .

Sad as it looks, we as a culture are boring ourselves to death because we are so reliant upon ideographic language. Indeed, we're overwrought by it. Why? Because for some reason our society has bought into the belief that this conceptual mode of communication is the only objective way of doing things. After all, there is no room for subjective emotions in the reporting of the factual events of the (objective) world now, is there? IS THERE? IS THERE EVER!

✿ ✿ ✿

BIO-CULTURE	MAIA	MYTHOS	R. BRAIN MIMETIC	L. BRAIN MIMETIC	LOGOS
BRAIN LOCATION	REPTILIAN	MAMMALIAN	R. HEMI. NEO-CORTEX	L. HEMI. NEO-CORTEX	L.HEMI. NEO-CORTEX
PRIMARY SENSE	KINES-THESIA	AURALITY	VISUAL IMAGISTIC	VISUAL CON-CEPTUAL	DIGITAL
DIVINE IS; PERCEIVED AS:	THE ONE; A FEELING	THE GODS; A HARMONY	DEMIURGE; AN IMAGE	SAVIOR; A CONCEPT	MACHINE; THE UNIFIED THEORY
RELIGIOUS SYSTEM	PANTHEISM	POLYTHEISM	MONO-THEISM	MONO-THEISM	ATHEISM AGNOSTIC
THE BODY IMAGE	SERIES OF ENERGY CIRCLES	A MUSICAL STRING	MICROCOSM OF NATURE	DUALITY OF BODY AND SOUL	MACHINE
METHOD OF OBTAINING KNOWLEDGE	BODY'S PHYSICAL RESPONSES	EXERCISE ALL HUMAN FACULTIES	GURU TEACHER MASTER	THE BOOK	THE LAW, UNIVERSAL FORMULA
SOCIAL SYSTEM	DIA-MATRIARCHY	GYNE-MATRIARCHY	DEUS-PATRIARCHY	ANTHRO-PATRIARCHY	ANTHRO-HIERARCHY
ESCHAT-OLOGY	RETURN TO THE ONE	MATRIX SURVIVES; LIFE FORCE BACK TO THE ONE	METEM-PSYCHOSIS OR APO-THEOSIS	DISSOLUTION OF BODY; SOUL IS IMMORTAL	NO PERSONAL SURVIVAL AFTER DEATH
LANGUAGE TYPE	MYTHO-POETIC	MYTHO-POETIC	MYTHO-POETIC	TRANS-ITIONAL	IDEO-GRAPHIC

The Biocultural Paradigm

Maia, Mythos, Mimesis and Logos: The Biocultural Paradigm

The acknowledgement of the existence of two distinct languages in the human mind carries an additional burden. That burden is similar to the problems one has in trying to understand someone speaking a different (ideographic) language from your own; like someone speaking English trying to understand someone speaking Spanish if the English speaker has no knowledge of the lingusitic rules operating in that other language, though both languages share a similar alphabetic system of notation. Similarly, the language of images, the mythopoetic language, has its own rules of "grammar" if you will. And although mythopoetic language may be expressed using the technologies of ideographic language, i.e., words functioning through some linguistic system, the meaning of those words is dependent upon entirely different mental functions. And the same way that similar alphabetic systems can yield different languages, the mythopoetic language yields divergent groups based on the primacy of a particular sense. And just as communities and cultures are formed by individuals who share common ideographic languages, the variation of mythopoetic languages likewise yields its own communities and cultures. And finally, despite this disparity, at the root, all the mythopoetic cultures organize their world by images and all the ideographic cultures base their world on concepts.

The biocultural paradigm[5] as summarized on the chart at the beginning of this chapter describes in detail the basic biocultural prototypes of human societies, using the criteria of both human languages as verified through our neurological and neurolinguistic functions. For the sake of clarity, the prototypes are linearly described so that the reader, at a glance might envision the development of each prototype based on the primacy of the neurological processes involved in its conception. This is not to contend that these types, presented as such, appear historically in such a hierarchical order of ascent or descent, for this notion of progressivism or historical development is only a concept contained in the ideographic language of our times; rather it is meant to bracket, for the sake of examination, each particular prototype as a whole, prior to exploring the ramifications of these types interacting both in the world and within the individual.

Of course it may seem odd to the reader that four out of five bioculture types thus noted are mythopoetic and only the fifth prototype, what I term the logos type is ideographic. This peculiar point shall be examined at length at the end of this chapter, for it is this factor upon which the thesis of this book hinges.

A simple explanation of this chart will prove first how these biocultural types are literally imprints, physical remnants of the mind processes that operate to create them, and also how these prototypes interrelate with one another, both culturally and correspondingly, biologically, and even cosmologically.

Beginning first with the maia prototype, one observes that kinesthesia is the sensory input that organizes this system. The origin of this function of kinesthesia is the hind brain, or reptilian brain, seat of our motory functions. Through the primacy of kinesthesia, a world is created by the invariant of rhythm; the churning of the seas, the dance of the god or goddess, the "one" breathing by self impulse. Correspondingly, the gyrations of sexual motion, the tactility of our nervous centers, the rhythmic, involuntary pounding of the heart, the flow of blood and the inhaling and exhaling of breath. What is a human? A whirling mass of energy circles, conditioned by the parameters of invariant patterns of movement. Pantheism is the religious expression of this prototype because the universe is motion and the One set it in motion, by the desire to move and we are all manifestations of the motion of the One and therefore we are all part of the One, and the One is therefore our "god." What happens when humans die? This is a ridiculous question. Perhaps the question should read, who dies? The answer, no one. Because in the Maia prototype the circles of energy that moved together for a while within "your" orbit (if you will) cause other motions and eventually regroup into other patterns in the same universe that is the One. Who is god? The One. How is god perceived? Breathe.

The mythos prototype is organized by the sensory criteria of sound. The neural connection within this prototype is the limbic system, seat of emotion. Just as our reptilian brain senses motion, our limbic brain senses motion, but through the interaction of waves, if you will. The limbic system then translates these waves in the body, thus forming a spacial existence, a world based on sound vibration. Each vibration carries a different act, a different perspective, a different feeling. The human body is literally a string that, depending upon how it is cut, divided, dismembered, creates a corresponding individual. Polythesism is the religious system of the mythos prototype because each god is one cut in the string that is the human, one act of the many acts that form the human. By training the body sensorium to attune itself, to learn to modulate in the face of stagnation, a person aspires to integrate his own song to the "harmony of the spheres," the universal harmony that is the gods and goddesses. What is the ultimate fate of humans? Either they attune themselves to the divine harmonia and become gods themselves or, their attunment is of a lesser kind, perhaps of the vibrational pattern of an animal, and they, upon the dissolution of the tension of their human string, reincarnate as that other image.[6] Understand that the belief system here is complex: Vibrational patterns manifest themselves in (what we call) material form. The form then, the body, is merely an IMAGE of the culmination of harmonies operating in a particular geometric space. To change the body, change the vibration. How do the gods manifest themselves? Feelings of consolation or desolation. Who guides you? Your daimon, your inner voice, or god center. And how do you know when you've achieved divine harmonia? When you can choose the best act, song, god, from among those possible at any given time.

The mimetic prototype is really the prototype that is the most complex, because of its capacity to integrate the other mythopoetic prototypes and also because of its capacity to reinterpret them and occasionally, misinterpret them in the course of integration. This prototype is called mimetic, because that is exactly what it does: it imitates. The locus of the right brain mimetic prototype in the brain is the right hemisphere of the neocortex. Of course this hemisphere is the traffic director of the previous two brain centers and it is also the informant to the left hemisphere of the raw data being processed, so that the left brain can decide whether or not to consciously override the data.[7] So, likewise, the mimetic protoculture, in an attempt to facilitate harmony among the parts, organizes itself around the criteria of sight, the most "democratic" of the senses. Vision is the wave that causes the necessary holographic interference with the vibration of the limbic and reptilian senses, causing the geometric image to "appear." Visible things are holograms of our own brain's interference patterns. And in the mimetic prototype, these "things" are the invariants that make up the world. And ourselves. What is the body? A cooperative ecosystem, a microcosm of nature. And who is god? The creative demiurge that puts things together out of nothing (no thing!) and gives them form, material independence, visual integrity. God is the great hologram maker: the traffic director, the interpreter of all things, the omniscient and omnipotent one. And there is only ONE GOD, though as we shall later see, he can sometimes be envisioned as three persons: the trinity.

Another function of the mimetic system of vision is that the creation of geometric forms gives rise to the emphasis on an objective world OUT THERE, separate and distinct from the maker, for the reptilian and limbic input do not originate from the right hemisphere, though they are "formed" there. And herein lie the ultimate cosmological problems for the mimetic biocultural prototype as well as the birth of a great schism within this mimetic prototype itself: What happened in the Beginning? Who are we? What survives our death? Where are we going? These questions are natural results of espousing this interpretation mode, but because of the unique construct of the right hemisphere, these questions have caused perhaps the most serious problems of the species. Why? Because of a very important shift in the brain: the shift from the right brain being the traffic director to the left brain being the dictator. You see (no pun intended), in a mimetic biocultural system both possibilities can exist: a god *could* be the One of maia made manifest in its multifarious forms through the incorporation of the mythic and mimetic vibrational patterns OR god is the left brain judge of humanity, outside the creation of the world, merely criticizing it, chastising it, and ultimately, punishing it. Laws and moral codes of behavior are manifestations of the left brain god. And these laws must be written proscriptively, do NOT x, do NOT y; because in order to be comprehended by the left brain, anything must first be created by the right brain and then overridden.[8] Mysticism and signs of the spirit are the manifestations of the right brained god. Why? Because the god of the right brain is the harmonious interaction of the parts toward the whole. It is one person in three forms. Three vibrational systems manifesting the hologram that is love.

Two answers, one system. And now, the real problem: what about the body. The mind/body split is a fiction created by the interference of the left brain in the mimetic right brained world. The body is evil when it cannot be controlled, when the left brained god of override is disobeyed.[9] It's evil because it literally has a mind of its own, and sometimes that mind, or more accurately, those minds perceive differently from their left brained interpreter. So who has the final say? It depends on the mimetic cultures' slant. Right brained mimetics say the body; for god's will can only be felt from the inside. If it comes from the outside, it is NOT god. Left brains say the mind of the self, the "true" immortal part of the humans, the soul. And what is this "immortal" self? The seat of the decision making process between these two interpretations of the world. And what is the soul? The "immortal" construct that the left brain conceptualizes using the holographic image made by the right brain in conjunction with the other brains. It is that conceptual reality that houses the continuity of that decision maker after the body is long gone. So what is the fate of humans in both mimetic systems? The spiritual resurrection of the biological *and*/or conceptual holographic "form." And where does it resurrect *to*? The only thing we know for sure is, not here.

Finally we come to the end of the mythopoetic models. With the birth of the logos biocultural type the individual moves out of the realm of the involuntary brains and relies solely on the conceptual systems invented by the left brain.[10] Here the world is constructed solely on the abstract logical systems that are the children of the left hemisphere of the neocortex. There is no more fight between body and mind; indeed there is no need for the input of the body whatsoever, since the left brain can construct its own substitution system in the concept. The concept is an entity void of motion, emotion, locomotion; therefore, it lacks vision because there is NO INTERFERENCE pattern to set up any. So, if it can be conceptualized, conceptualize! Technological advances occur rapidly in logos protocultural society because they are invented to replicate the acts of the left brain without having to rely on the instrumentation of the human body. One only needs a left hemisphere to survive in this culture. This mind then is apprehended as a theoretical invariant designed to serve as observer of everything that is not itself, which is everything. It perceives this way because that is how the left brain receives information from the right brain, but the conscious mind doesn't consciously know it because this right brain input is imprinted in the body using a language of images which this logos brain DOES NOT UNDERSTAND. It first, in order to bring this data into "consciousness" (into a conceptual form) must translate the images. In the translation process, since the left brain can only understand concepts, it appears that the source of the data is "out there". The translation process really occurs in the body, but since the body is not part of the system, it becomes the invariant form, the instrumentation machine that perceives, or discovers, a seemingly "objective" reality. But since it cannot do this directly, it must first create a logical system through which to perceive such that what is viewed in the image mode is given its own form, albeit not holographic, but conceptual. Hence, the birth of the ideographic mode: alphabetic systems, logical systems, theory, ideology, theology, science . . . So what is the ultimate goal? To know everything "out there." How? To investigate everything out there, testing it, testing how we tested it to make sure we didn't make a mistake in the conceptual process of testing it. And who is god? No, what is god? The ultimate theory that answers all of our questions. And who lives and who dies in this universe? Everything but "theory." For all conception is "stillborn."

✿ ✿ ✿

The Triune Brain and the Biocultural Paradigm

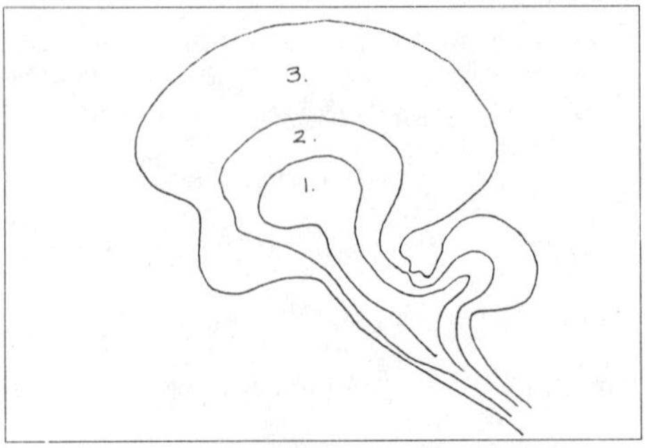

The Triune Brain (after P. MacLean)

Chart Showing the Brain Centers and Behavioral Traits by Bioculture:

Brain Center	**Bioculture**	**Primary Sense**	**Behavioral Traits**
Reptilian Limbic (amygdala) Limbic (septal)	MAIA	kinesthesia, olfactory	involuntary body functions, motor functions, eating, sexual funtions, territorialism, ritualism
Limbic (prefrontal cortex)	MYTHOS	auditory, visual	internalization, empathy, bodily homeostatis
Neocortex (R. Hemisphere)	RIGHT BRAIN MIMETIC	imagistic visual	visual sexual-social criteria, "self" development, externalization
Neocortex (L. Hemisphere)	LEFT BRAIN MIMETIC	conceptual visual	symbolic language system, causality, beliefs, extern-lization of body phenomena
Neocortex L. Hemisphere (Interpreter)	LOGOS	a-sensory digital	opinion, inference, seat of logical fallacies, conceptual systems with no basis in body of nature, extern-lization of its own mental functions outside iteself

The Scientific Basis of the Biocultural Paradigm

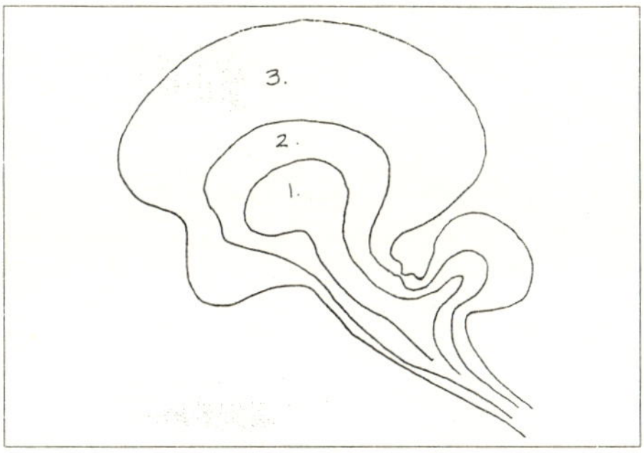

The Triune Brain: The Evolutionary Development of our Many Minds

Paul D. MacLean, M.D., neurobiological and behavioral researcher, developed the concept of the triune brain, based on the evolutionary development of the three formations comprising the human brain: the Reptilian formation, the Paleomammalian region and the Neomammalian area. (See figure 1.) According to MacLean, these three "brains" contain, "radically different chemistry and structure, and (are) in an evolutionary sense countless generations apart. (What) we have, so to speak (is) a hierarchy of three brains in one . . . a triune brain[11].

Thus, though each evolutionary layer of the human brain builds upon the earlier formations, we, as humans, maintain and continue to utilize these earlier systems and they do, in fact, comprise our present brains. The most stunning hitch to this triune brain structure, though, for purposes of the biocultural paradigm, is that, "what seems to be notably lacking is a neural code for intersignaling IN VERBAL TERMS" (emphasis added) among these brains[12]. And this hitch, raises the neurobiological issues to biocultural ones.

The Reptilian "brain," the one we humans "share" with reptiles, appears to be responsible for our intrinsic forces of self preservation and species preservation in general. In addition to these basic functions, MacLean has suggested select patterns of behavior evident in reptilian communities which are also evident in mammalian and human communities as well. These functions include: territorialism, ritualistic displays of territorial defense, routinization of activities, hunting, foraging, hoarding, formation of social groups, grooming, courtship, mating and breeding[13]. Specifically, those tendencies of lower forms that are clearly indicated in humans, MacLean identifies as isopraxis (imitation), preservative, re-enactment, tropistic and deceptive; in short, a wide array of instinctual proclivities, leading to the conclusion that the Reptilian complex does indeed regulate a large portion of our bodies' instinctive behaviors[14].

The Paleomammalian "brain" with its connections to the brain stem make up the limbic system. According to MacLean, the limbic system is the source of emotional behaviors and experiences, the most evident of which is the behavior of parental care. The limbic system is divided into three basic areas, according to MacLean, each of which refer to different aspects of our mammalian tendencies. The first area, connected with the amygdala, is concerned with feeding, fighting and self-protection. The second area, in the septal region, concerns genital and procreative functions. The third region, tied to the pre-frontal cortex, allows us to obtain "insight" and therefore serves as the "empathetic link" to the neo-cortex, and perhaps to our capacity for compassion[15]. It is also important to note that the first two areas comprising the limbic system are related to the olfactory regions of the brain while the third area bypasses this region . MacLean sees this as being the source of vision superseding olfactory sensation in our socio-sexual behavior[16].

Another curious aspect of our limbic system is the manner in which it apprehends data: by the auditory criteria of wave interference patterns. Joseph Chilton Pearce describes this unique phenomenon as follows:

> Our limbic system resonates with waves just as the Reptilian system does with particles. With the help of the neocortex, our limbic system can translate waves directly, before their display as particles is picked up by the R-system . . . The report from our body gives tangibility to our felt-space just as the feedback from the eye concretizes an otherwise incomplete visual system[17].

To conclude then, the limbic system, as it is linked to the Reptilian brain, controls such internally instinctive aspects as feeding, fighting, self protection, and genital procreative functions. The limbic system, as it relates to the neocortex, appears to control our capacity for empathy (that is, internal identification with another), and perhaps even our internal sense of self. It also aids in our capacity, (through its apprehension of modulating waves within the body,) to create the geometric, three dimensional "forms" out of which our material world later "appears".

Finally, the neo-mammalian brain, as identified by Paul MacLean is that region of our brain "oriented primarily to the external world and seems to serve as a kind of problem solving and memorizing device ... With its focus on material things, the neocortex develops somewhat like a coldly reasoning, heartless computer . . . that has the capacity to devise the most violent ways of destroying our own kind as well as other forms of life."[18] (This is of course, a description of the neocortex without the inclusion of the prefrontal cortex, which of course mitigates this interpretation.)

To sum up, then the triune brain and its functions:

> Our Reptilian system . . . dominates and orders lower ganglia into its service . . . The limbic system orders those physical instincts into a . . . higher order of operations as will be needed by the neocortex. The unifying right hemisphere (operating in conjunction with the left) utilizes the limbic structure for intuition and imagination. The left hemisphere (eventually) becomes dominant over and employs the right hemisphere for concrete operations[19].

With the evolutionary development thus concluded, we shall now turn our attention to the most recent developments in the studies of the cerebral hemispheres of the neo-cortex and their ramifications .

✿ ✿ ✿

The Cerebral Hemispheres: "Right Brain/Left Brain"

Much popular literature has been written on right brain/left brain specialization and other related issues in psychology. Though the preponderance of evidence today seems to indicate that there is more interactive than distinctive functioning between these hemispheres (not to mention the role of the corpus callosum as related to these issues); there are certain distinctions that are indeed worth noting, for our discussion. For example, the findings of the "split brain" research performed by Sperry and Gazzaniga, Gazzaniga's discovery of the "interpreter module" in the left hemisphere, and finally, the proof that Selection Theory has application in human neurobiological development, each provide some biological foundation for the veracity of looking at cultures, not as ethnically based organizations, but rather as bioculturally based organizations.

Of all the findings articulated by the split brain research performed by Sperry and Gazzaniga, the one that is most intriguing for the biocultural paradigm is the discovery that that old adage could be true that the right hand doesn't always know what the left hand is doing. To quote Gazzaniga:

> In short, we had collected data that revealed that even within the normal brain the right hemisphere possessed knowledge that was encoded in a form that remained insulated from the search process of the left hemisphere's language system.20

So what are these diverse articulations of the right hemisphere that seem to be so elusive to our human language system? Further, how do we access this hidden knowledge, and recognize it, if not through the linguistic channels of the left hemisphere? And most importantly, is it possible that if our left hemisphere cannot recognize this knowledge, it will assert (to itself) that this knowledge simply does not exist? The answers to these questions form the foundation of our biocultural paradigm's claim to authenticity.

The conclusions drawn from split brain research and other such studies regarding the unique features of the right hemisphere of the neo-cortex are that it performs better with spacial functions, visualization (that is, remembering images), "Gestalt-type" processing, and non-verbal stimuli, including extensive auditory discrimination. This is not to say, however, that the right hemisphere is devoid of verbal comprehension; for in fact, it is the right hemisphere that seems to be responsible for a person's appreciation of the sophisticated nuances of language, such as puns. (Recent studies have shown that individuals with right hemispheric damage are unable to distinguish linguistic subtleties; rather, they simply internalize the superficial meanings of the linguistic modifiers as perceived. In other words, they just don't have any senses of humor![21]) What the right hemisphere cannot do though, is abstract causal relationships between conceptual ideas. In other words, it cannot interpret.

Conversely, the left hemisphere is responsible for making causal inferences, for apprehending complex linguistic constructions (symbolic), for learning grammar, and for voluntary facial expression. (So this is the hemisphere that allows us to "fake a smile.")

The relationship of the left hemisphere to the other brains is best described by Pearce in the following passage:

> The left hemisphere, largely isolated from the limbic system, uses its connection with the right hemisphere to draw information from, and communicate its abstract ideas to the rest of the system. The routine is Left hemisphere to Right hemisphere to limbic to Reptilian . . . Through its possibility for isolation, the left brain's lateral actions can be independent and NOT subject to the general rules governing the integrated brains. This means the left hemisphere doesn't have to answer to the organic intelligence governing life as a whole, as the lower structures do.[22]

And this is due to the unique apparatus in the left hemisphere that contains what Gazzaniga calls the "interpreter".23

The role of the interpreter, according to Gazzaniga is to "allow for thought about the implications of actions, generated both by others as well as the self . . . (and to) grasp a social context and its meaning for personal survival."[24] In other words, the interpreter allows us humans to infer causality and to have beliefs. But, as several cases have shown, there are certain times when this interpreter forces interpretation on things which in fact have neither causal relationship nor believability. For example, in one of Gazzaniga's split brain studies a patient was shown a snow scene to the right hemisphere and a chicken claw to the left hemisphere. When asked to associate these images with a series of pictures placed before him, he correctly chose a chicken for the right and a shovel for the left. (In other words, the right hemisphere correctly analyzed and chose the response.) When the individual was required to articulate his reasons for choosing these items, he (his left hemisphere) responded, "Oh, that's simple. The chicken claw goes with the chicken and you need a shovel to clean out the chicken shed."[25] In other words, the left hemisphere CONSTRUCTED from its limited knowledge, an interpretation of what IT had experienced WITHOUT consideration about the knowledge of the snow scene: information that was only internalized by the right hemisphere and therefore unarticulated by the scenario of the left hemisphere. And this leads us to the final issue regarding the role of these two hemispheres: Which one of them is the boss? The answer it seems, is the left hemisphere BECAUSE of the role of this interpreter. As Pearce notes:

> With an abstract, semantic language we can create thoughts "out of the blue," thoughts that arise from the act of thinking, rather than from responses to objects themselves . . . We can extract information out of context and operate on it with self-created ideas that are discrete and self referring rather than balanced within a larger frame of reference.[26]

That is not to say, however, that we humans are genetically determined solely by the hard wired hemispheric functioning of our brains; but we are, to a certain extent more a product of nature than we are of nurture. Selectionist theory seems to indicate that as much as the environment may play a role in one's choosing particular propensities over others, at the root of it all, there are certain pre-existing capacities that basically determine the conditioned DIRECTION of our propensities. "For the selectionist," writes Gazzaniga, "the absolute truth is that all we do in life is discover what is already built into our brains . . . The environment selects from the built-in options; it does not modify them."[27] So, with this as a backdrop then, what role, if any, does acculturation play in civilization? It seems that for the most part, the windows for us humans to make significant genetic "choices" that would eventually manifest themselves in the cultural milieu is very, very narrow.

Joseph Chilton Pearce, in his book, *Evolution's End*, traces the neural development of children in an attempt to thwart what he sees as the species' unwitting decline from its spiralling evolutionary ascent. His research, though not directed towards the presupposition of a biocultural paradigm, does in fact substantiate the biological origins and reasons why certain biocultures develop as they do. For example, in his discussion of brain growth spurts in children, and their responsiveness in development, we see that the Reptilian Brain begins in utero and continues developing to age seven. The Limbic system begins development at age one and ends at age eleven. The right hemisphere of the neocortex develops at age four and becomes conditioned by age fifteen and the left hemisphere begins to develop at age seven and ends sometime in early adulthood. The implications of these findings are significant because if we, as humans "miss" the windows of malleability as they refer to each of our brains, those capacities will be lost forever once the "selectivity" phase is over. In short, use it or lose it does seem to be the message here.[28]

So what of the correlation between biology and culture? It appears rather certain that indeed the sins of the father are burdened upon the son, simply because of the tendency among indigenous social groups to repeat what was done to them in the next generation. We know now, however, that it is because both nature and nurture share a common window; a window that can only be opened through multi-(bio)cultural integration of the young. And this can only be achieved through education by the criteria of all the biocultures, which would simultaneously achieve biological reintegration of ourselves as a species.

CHAPTER TWO

CREATION: FROM MYTH TO IDEOLOGY

The Mythopoetic Image

Let us begin this chapter with a few linguistic clarifications. First, it must be understood that mythopoetic language is NOT a language of symbols. Symbols are part of the ideographic representational system created by the left brain as a substitute for the image holographically made by the right brain. Neither is mythopoetic language a language of metaphors; for metaphors are likewise ideographic place holders marking conceptual realities. In fact, a mythopoetic image is an act in form; in flowing form. All of these other interpretations of images betray the fact that their origin lies not in the act of image-making, but rather in talk about images MADE; about a thing out there: dog, dragon, cow. It is a mistake, therefore to approach any work using the criteria of mythopoetic language with these linguistic presuppositions. So, how does one write about the act of image making? In other words, what are the written remnants of mythopoetic language if not symbols or metaphors? Curiously enough, myths are. To provide a suitable, albeit not entirely correct analogy, one might say that as nouns are to ideographic language, verbs are to mythopoetic language. In other words, if you imagine myths to be written accounts entirely composed of verbs; not written verbs, but the actions of the verbs themselves, you will come close to understanding how to "read" mythopoetic language. And in myths of Creation, the "verbs" that are being described are the "how's" and the "why's" and the "what can you expect to find's" of the acts involved in making images. They do not distinguish how to make images of what; they distinguish what images result from what actions. And in all creation myths the four acts of maia, mythos, mimesis and logos are manifested as images in the myths; but their purpose is to reveal first the "history" of the act of imagining and then to relate which of these four acts has primacy in this myth system (meaning in the myth that the teller is about to tell at that moment.) Can we as modern readers, pick out these images in the myths? Of course we can. The problem occurs not when we identify the images, but when we presuppose that the way in which we "see" the image is the same way in which the image is to be interpreted in this language system. What we read are these images as symbols; what they are are memory points to retrace the act. Big difference.

To further complicate things, sometimes the images of one mythopoetic biocultural group are appropriated AS SYMBOLS for another group and interpreted mimetically to create a seemingly new myth, when in fact, this new myth is not a myth at all but a hybrid creation of mythopoetic form with ideographic intention. This usually occurs in cultures following what we call the left brain mimetic biocultural paradigm. These new myths are really the vestiges of the origin of ideographic language; they are the linguistic archaeological evidence of the neurological shift in the culture from the aforementioned right brain traffic director to left brain dictator. These "myths" are fascinating because they occur in all myth systems, precisely to attempt to mark the birth of this new language in the brain, but without the wherewithal of that new language, to describe it.

What follows then is a concrete analysis first of mythopoetic creation as evidenced in the actual creation myths of several mythopoetic biocultures. Then, an examination of several of these "new" myths, which uncover the shift from mythopoetic language to ideographic language and finally, to the modes in which ideographic language interprets the issue of creation; not as myths at all, but as theories about creation, or worse, as formulae about creation! Of course, what these ideographic accounts are accounting FOR are the acts of conceptualizing, so in a way, even their creation theories are "myths"; for they are telling the story of the acts that the brain doing the telling is doing. And that brain is making theories; therefore, creating becomes a theory of making a theory of creating.

✿ ✿ ✿

Mythopoetic Creation and Its Varieties

At the outset of our discussion of the biocultural paradigm, we distinguished three biocultural prototypes (maia, mythos and mimesis), corresponding to the reptilian, limbic and right hemisphere of the neo-cortex, by their choice of the primary sense in which they organize their world. Thus the maia prototype emphasizes kinesthesia; mythos uses sound and mimesis, sight. What links these three prototypes, however, is that all three rely primarily on mythopoetic language as their form of communication. Therefore, in analyzing the myth systems of these groups one will find that they all share similar myths of creation, since these myths are in fact a description of the criteria for using mythopoetic language. The discovery of what biocultural group one is dealing with then comes only when we realize which of these senses is primary in that group. And sometimes, even the faculty of thinking (conceptualizing in the ideographic mode) is described in myths of creation because some mythopoetic bioculture types (usually the maia type) separated conceptualizing from imagining, and viewed this act of thinking merely as one more human sense, thus it is accounted for in the creation myth. So the analysis you are about to witness is an attempt to ideographically translate the mythopoetic language of creation myths. (No wonder my students had so much trouble writing essays for me!)

Myths of creation in general first relate the invariants: I'll call them chaos and cosmos; loosely translated, what was before I started image making using this sense and what happened when this sense started to move and image making began. Chaos can best be described as undifferentiated possibility; cosmos as the first rumblings of imagining and what occurs thereafter. Are there images that we, as ideographic readers can identify and say, here's chaos? Yes. What are they? It depends. Sometimes, a dragon or snake describes chaos. Sometimes, the churning waters. In other cases, chaos is a geometric form, a circle usually; and other times, chaos is a goddess or the father of a new god, or a hero. And sometimes, chaos is merely the one breathing by its own impulse.[1] How can you tell when these images are chaos? Look at what they DO in the myth.

I have reprinted in the appendix to this chapter, several creation myths from various cultures. Let's tackle them now to try and better understand how to "read" this new (old) language. For if we continue talking about them, we fall again into the ideographic mire of conceptualizing; an exercise which defeats the purpose of this chapter. Please refer to the numbered samples in the appendix to this chapter while reading the narrative that follows.

Creation in Four Moves

All mythopoetic creation myths, whether from a maia, mythos or right brain mimetic bioculture, include descriptions of the four mental properties necessary to bring forth a manifested world FROM THE INSIDE.[2] These descriptions are reflected in certain, distinguishable images. The primacy of the sensation of the bioculture, i.e., kinesthesia for maia, is determined by the force that is responsible, within the myth itself, for bringing about cosmos from the primordial chaos. Chaos, then, in mythopoetic terms is defined as undifferentiated possibility, a dragon, a mixture of elements not yet formed, the undivided One, the universal goddess; while cosmos, is the impetus for differentiation. Thus, it is only when something begins to happen in a creation myth that chaos takes on a mythopoetic appearance, a form. But note also that whenever chaos is thus viewed, there is always someone acting upon that form, slaying, cutting, dismembering that image. Or, as in the case of the goddess as chaos, note that her formless form nevertheless gives rise to all forms. So, to conclude here, chaos is either potentiality "no-thing" or the source from where all things COME. BUT, it is only the second, when one is referring to it in relation to some other act. That act, would be the act of creating COSMOS; the act of dismembering the unity of the origin in order to make new worlds. Technically speaking, cosmos is the human act of dismembering the senses by the act of imagining[3] (See appendix on how this is done, specifically.) And in myth, this act is always portrayed as anything that breaks open by cutting apart, or slaying or castrating whatever image provided the unity of sensation in the person. So, if I am a maia person, and my unity is the sense of touch, then chaos is inaction and cosmos is action. If I am a mythos person, and my unity is sound, then chaos is silence and cosmos is sound. Likewise, if I am a mimetic person and my unity is sight, then chaos is what there was before things I saw, or darkness and cosmos is the world of things. And an extension of this understanding sheds light on the corresponding bioculture types. Deity in the culture types are whatever appear as the saviors of the chaos of that system. Thus, in a maia bioculture, the cycle of movement (Hindu Rta, for example) is the savior; in a mythos culture, the gods are the saviors; in a mimetic culture, either the demiurge god (right brain) OR a hero (left brain) is the savior. And this is a UNIVERSAL phenomenon.

If we take the above as a model, then we can best understand the next point. Myths of creation, since they report the act of image making, account for the birth of ALL three senses (touch, sound, sight) because all three TOGETHER create the world of images, the "material world." So in myths of creation, there is a corresponding image to each of the above shifts from chaos to cosmos (that is, as any individual goes on creating the image by applying the different senses to making it) and these images are universal in mythopoetic language.

One can catalog these images as follows:

1. Inaction is the chaos of touch; the dragon or snake (later, monster) is the cosmos of touch.
2. This same dragon is the chaos of sound and the gods are the cosmos of sound.
3. The demiurge god, (or Chief god) is the chaos of sight, and the hero is the cosmos of sight. (The hero is half god/half human).
4. The repetition of this cycle by each human yields the creation of the entire cosmos, because with the sense of sight, the world APPEARS. And why does it appear? Because of humans, literally. This is why creation myths end with the creation of humans.

To recap then, by the act of imagining, which is the act of dismembering the human sensorium in such a way as to re-organize the original unity, creation occurs.

Step One: Kinesthesia

The first act of creation is movement. (Incidentally, these myths never answer the question, who began the movement. And frankly, no one even cares, because all that counts is that we're here now. More about this later.) Wherever you are, in order to create ANYTHING, first you must MOVE. In myths of creation, each bioculture represents this a different way, but in maia biocultures in particular, it is through this movement that cosmos is made manifest. (See Appendix 2.4, Sample One).

Step Two: Aurality through Emotion

The central focus of aurality is that one can only create reality within the sensorium of sound through the mediation of at least two separate tones; otherwise, one is stuck in chaos (the monotonous drone of unity). With this in mind, the separation or differentiation of opposites is what constitutes the ordering out of chaos in this, mythos phase of creation. Within the space between the two differentiating tones, a world of harmonics is born, the mythic world of cosmos. The images of this second phase of creation are as follows. Chaos is the uncut, undifferentiated string (or unitone); the opposites (male and female creator deities) represent the forces which together, through a sacred union and then separation, bring forth cosmos.Cosmos is mythopoetically perceived as the polytheistic deities that spring from the sacred union. (See Appendix 2.4, Sample Two.)

Step Three: The Creation of Geometric Form

The conflicting vibrational patterns of touch and sound in the imaginer cause a visible sensation to appear. In myths of the right brain mimetic biocultures, the invariant of the imaginer now takes on the role of demiurge; the creator of forms out of geometric possibilities. In myths of right brain mimesis,this demiurge is viewed as a creator god who fashions the world of "Things" (out of what, is not known nor even relevant). In short, the demiurge is the right brain, which, taking the sensory data from the lower brains, forms plausible scenarios out of which to make reality. So, Chaos is the world before forms, and cosmos is the world "formed." (See Appendix 2.4, Sample Three.)

Step Four: The Creation of Material Form

From the imaginative realm to the material realm, the world moves from right to left brain mimesis, for in order for there to be material creation, a conceptual copy must be created as blueprint, if you will. Correspondingly, in these myths, the characters who release cosmos out of the bonds of chaos are the heroes: the physical manifestations of a deity in material form. (See Appendix 2.4, Sample Four.)

Conclusion:

As you can see from above, each of the levels of creation is mentioned in all myths of creation, but the emphasis, by bioculture is determined by who or what brings that particular creation out of chaos and into cosmos. In maia biocultures, the myths begin with kinesthesia. This impetus sets the senses in motion and through the conscious effort of dismembering the unity of the one that before didn't move (to us, the one doing the creating) by the successive acts of shifting the senses, regeneration, rejuvenation and even immortality is gained. And all of this happens in this world by humans! We control the invariant structures of our biological makeup by keeping them in motion and this is how we make, or break the universe. This is the message of mythopoetic creation, and no other.

✿ ✿ ✿

Interlude: Myths of the Fall and Human Redemption

As we previously discussed, in myths of creation, gods and heroes are the cosmos of sound and are therefore the chaos of sight. In other words, in mythopoetic language, the gods are the artificers of the visible world, since they bring it out of the chaos of darkness. Their creation, then, is the visible world, and also, humanity. In mythopoetic language, these demiurge gods are often described as architects or great geometers, as the following sample from Ovid shows:

> And then some god, a higher nature,
> Stopped this controversy and cut
> The sky from air, and the sea from the land,
> And the thickened air was distinguished from clear . . .
> And after he had divided that heap
> And put it back into proper order,
> He bound each part tight into place
> Forming a peaceful harmony . . .[4]

It is these gods who are also responsible for making humans. And how are humans made? Usually from a combination of all the earthly elements (earth, air, fire and water) plus some element of the demiurge himself/herself. In a maia bioculture, the element is desire (the force that impels); in a mythos culture it is the creation of the archetypal woman, specifically Pandora ("all gifts"), as she supplies the womb as matrix for geometric form that carries all the archetypal acts noble or ignoble within herself; and in a mimetic culture, it is the image of the deity (the actual geometric form) that the demiurge adds to his creation. This element that he/she adds is usually what gives the reader some indication of what bioculture is operating in the particular myth of creation, for only here do these mythopoetic biocultures really distinguish themselves. So, for example, now we can see that the Rg Veda is a document of a maia bioculture: ("The immortal has a common origin with the mortal, and moves up and down by its power." 1.164, 38)[5] . The best example of a mythos bioculture might be the Greeks of the 7th c. B.C. ("And Hephaestus and Athena molded a modest maiden . . . and called her Pandora, because all who have their homes on Olympus each gave her a gift.")[6] And Ovid's work, *The Metamorphoses*, above is a remnant of a right brain mimetic bioculture.

With the creation of man, most creation myths end, but not before a discussion about "the fall." The myths of the fall are the recounting of the ultimate "sin" a human within that culture type might perform. And in each of these particular bioculture types, the fall of man is significant, and although each type expresses different reasons for the fall, they share, mythopoetically a single rationale. Mankind's fall occurs when stagnation of the senses occurs. Stagnation causes the return to chaos, but not the chaos of creation; the chaos of destruction. And when humans are destroyed, there is NO MORE HOPE FOR RECREATION. So, in maia cultures, movement is primary and although one can make gods; no god is exclusive. Everything must keep moving. And in mythos cultures, harmonia is primary and no one act or perspective can take priority over the others. So if you choose to favor one god over the others, you will be destroyed because you have not allowed the others to also live. And now we once again come to the problem culture: the mimetic. What is the fall of man here? Actually, there are two versions: a right brained version and a left brained version, though they are both derived from the same two basic myths. Because the mimetic group offended their creator(s), either:

a. The visible world is destroyed by a universal flood or a universal fire. Some couple is always saved to re-create a new world and a new race of humans (The right brained model) OR

b. The group is expelled from the company of the gods and their abode. (The left brained model.)

And how does a human, once "fallen" get back on the right track? In other words, what are the paths of redemption in these models? Sacrifice.[7] What IS sacrifice? In maia biocultures: giving into the aperspectival flow of life (Soma Sacrifice); in mythos, the ultimate modulation of physical and mental form (hence, the athletic games etc.); and in mimetic biocultures? Well, guess what — it depends. In right brain mimetic cultures it is the acknowledgement that you are not a fixed entity, nor is the world; therefore sacrifice is the detachment from the fruits of your action. But in a left brained bioculture, sacrifice is RENUNCIATION OF THE ORIGIN OF YOUR ACTIONS: YOUR BODY.

❁ ❁ ❁

The Left Brain and the Mind/Body Split

The notion that the body is the source of evil in humans has a very rich history; one that originates, oddly enough in left brain mimetic creation myths of the fall. The rudimentary biological explanation is as follows. We have shown how the reptilian and limbic brains, under the organization of the right side of the neocortex, form an information center in the human body to account for its survival. The function of the left hemisphere of the neocortex is to take this organized data and interpret it to itself. Sometimes, it allows the data to stay in its original form, as processed and sometimes it overrides the incoming data, thereby inhibiting or transforming it into something else. When the left brain does something with the information, it must first TRANSLATE it into the language of the left brain; into ideographic form. Then, whatever is stored in memory is stored twice; once as a series of acts in the body and once as organized and conceptualized information in the left hemisphere. Keeping this in mind then, let us turn our attention to a very famous myth of the fall; that is, the one in Genesis. For in this myth, the birth of the left brain, the origin of the mind/body split and even the creation of ideographic thinking are accounted for and described.

Before venturing into new realms, let us first explore the myth of the fall in Genesis[8] using the criteria set forth in the preceding section of this chapter; let us analyze it for the images that would have to be there if the myth were indeed a vestige of a mythopoetic bioculture. Who is the deity? A creator god; therefore, it is certainly the work of a mimetic bioculture. What element does man share with its deity? The IMAGE of the deity. This is another clue that we are dealing with a mimetic bioculture. And finally, to determine if it truly contains mythopoetic language, are there any other images in the myth that mark the acts of the other senses besides sight? In other words, are the images of maia and mythos also here? Surprisingly, YES. But WHAT are they doing here? Alas, the big question. Let's look at the images in the myth of the fall to determine the answer to this question:

a serpent

a creator god

a man made by god

a woman made by god ripping open the side of the man

two trees

some fruit

desire (on the part of Eve and then Adam)

Now, where do they fit in our mythopoetic language?

a serpent:	the cosmos of touch, remember?
a creator god:	the chaos of sight, remember?
the ripping open of Adam:	the cosmos of sound, remember?
the woman:	element the demiurge makes in mythos biotype
man in the image of god:	what demiurge makes in mimetic bio type
desire:	what demiurge makes in maia biotype
two trees:	(good and evil/life)?????
some fruit:	??????

So how are we to interpret this myth? Look at the action:

The creator god made the visible world and man, Adam. (creation by sight).

He rips open Adam's side and creates Eve. (sort of creation by sound.)

CREATOR GOD SETS LIMITS ON THE WORLD BY RESTRICTING THEIR ACCESS SAYING DO NOT EAT OF THE TREE OF THE KNOWLEDGE OF GOOD AND EVIL. THIS TREE IS IN THE CENTER OF EDEN. (This has no mythopoetic origin.)

Desire in Eve caused her to listen to the suggestion of the

serpent and eat of the fruit of the tree. (creation by maia, desire).

But what about the most significant objects in the myth: the "forbidden tree" and its fruit? Are these mythopoetic images? The answer, no. The trees and the fruit are not mythopoetic images; they are mythopoetic SYMBOLS and they refer to another type of creation which we thus far have not considered: procreation. The tree and the fruit represent the phallic and yonic components of human creation AND in some mimetic bioculture types, "myths" are created about human procreation and immortality; about how the human can "live forever." The human thus described; actually the bioculture that needs to account for the survival of this human, is always one which marks the neural act of the left brained interloper into the paradisal "natural" world. And once the human is CONSCIOUS of him/herself, i.e., once the left brain starts interfering with the "natural" flow of life as apprehended in the harmonious flow among the other brains with the body through which it flows, the trouble begins. First, humans are segregated not only from "paradise" but also from the part of themselves that manifests IN paradise, their human body. And once the body links with the world, the human is caught in an inextricable fight between the left brain observer and the world observed (including here the actions of their own brains in their own bodies). Further evidence in this myth itself that verifies this interpretation is the role of the god's proscription. DO NOT is a very odd series of words. In order to DO NOT, humans must first have to create the do and then cancel it. In other words, we are not biologically coded for DO NOT anywhere in the body EXCEPT in the left brain[9], the one that says YES or NO to the sensory input being delivered to the body by the other brains. It is this brain that creates consciousness as consciousness OF, because the data is perceived as a totality here, not as a series of dismembered senses out of which one forms a holographic reality. This myth does in fact contain the solution to this dilemma of the mind/body split very explicitly within it; unfortunately though, few from the culture reading it can understand the solution because of the misinterpretation of the story by the left brained interloper.

These two elements in the story of Adam and Eve, the tree and fruit plus the proscriptive voice of the creator god represent the first inklings that something else was going on in this story that does not exactly fit the model of mythopoetic creation, though some of the vestiges of that type of creation are there. This "myth," then is a mythopoetic representation of how to make the world AND it is an ideographic representation of procreation (the birth of the body) and simultaneously conceptualization (the birth of the independent left brain.) It accounts for biological creation through the phallic tree and yonic fruit SYMBOLS; and the BIRTH OF THE CONSCIOUSNESS of the left brain in the proscriptive voice of god.

Now, let's look at the myth again, with ideographic eyes, to see how the earlier mythic images have become transformed (by the left brain) into a story about how the left brain controls the body and how the body, when it disobeys is expelled. So how does one in this new mindset now, redeem himself? By following the rules of the left brain and renouncing to the carnal desires of the body. Isn't that the message?

A final problem: who created Eve? Depending on the answer to this question, a culture falls either into the mimetic, mythopoetic realm or, it jumps right out of the mimetic and right into the ideographic world of logos. If the group says god created her, then the group is mimetic. But if the group says Adam created her, since she was born of the SUBSTANCE of him, then we are dealing with a logos culture, and the story is no longer within the mythopoetic realm at all. It has already become a theory.

❁ ❁ ❁

Heroes to Saviors and the Mind/Mind Split

In mythopoetic language, the creation of thought is merely one more sense and myths of the hero recount the birth of this sense. The capacity to conceptualize gives rise to communities of individuals forming complex societies, new levels of communication through which these societies function, and the rise of technological advances which enable these societies to flourish. Unfortunately, this same capacity also gives rise to conscious patterns of segregation based upon hierarchical structures imposed by thought, written communication systems based on alphabetic substitutions and the like, and as a result, two distinct types of humans living within those societies: those whose primary language continues to be mythopoetic and those whose primary language shifts to ideographic. These two groups can usually co-exist in society because these two languages co-exist within the individual. But what happens when the society becomes overwrought by ideography at the expense of mythopoesis? Myths of the hero will tell you.

Recalling their origins from our mythopoetic model of creation, heroes are "born" in what we might term the right brained mimetic portion of creation. These humans were the ones who did not succumb to the "fall" and as a result they, by their example lead humans out of the mind/body morass and back to their "divine" origins: back to mythopoetic creation. So, naturally, heroes share some special attributes. They are not born "naturally"; their birth could not have been procreative because they are outside of the "fallen" human (read, left brained) model. So, the myths account for this by making their mothers royal virgins and their fathers, gods. Of course this is the case because they are visible manifestations, so their origin is in mimesis; thus, their mortal mother, but although they are human, they are not of human "substance" if you will. They were not, in other words, conceived (conceptually made, like other humans) for their origin is in the right brain, not the conceptual realm of the left brain; thus their father cannot be mortal: he must be the originator of mimetic creation: a god.

The "life" of a hero is not easy. He usually suffers tremendously, slaying monsters, performing superhuman feats and always overcoming DEATH. Then in the end, the gods reward him by making him one of them. A human raised up to become immortal. And because of the race of heroes, in myth, the world is saved; the project of creation is saved, because HUMANS are saved.

This is the basic story, from the mythopoetic cultures. Every time humanity has become so fallen into the illusion of ideography that annihilation is imminent, a hero is born to save the plan of creation by reminding them of their origins and their responsibilities. And every human who understands mythopoetic language can read these images in myths of the hero, and save their society and themselves. BUT how are these myths interpreted by the ideographic readers? Very differently.

First, understand that the origin of the logos bioculture is in the mimetic bioculture; it is merely created by a misreading of the images of mythopoetic language by ideographic language. The story is pretty similar; the MEANING is not. To backtrack for a moment then, the original fall of mimetic bioculture types is evident in the mind/body split, as discussed above. To recap, the left brain consciousness usurps the primacy of body consciousness and creates a separate and distinct entity in this human (a self) which in turn, by the replication of the left brain functions on itself, creates a schism within the human between body and mind. Body is perceived as outside the realm of mind and therefore as the source of evil (as we elaborated on in the previous section.) Once this habitual left brained system is internalized, a new language is formed in the brain; that is, a language of substitution to describe the holographic images produced by the right brain within the body, which, to the left brain are perceived as totalities outside itself, because the left brain is NOT the origin of these images thus wholly formed. In its exercise of interpreting this stagnant data, then, devoid of communication with the body, a substitution system of concepts is created in order that the left brain can "talk about" what it is being fed by the right brain without having to rely on the right brain and accordingly, the body. So, how is this human to be saved? Well, the body has to go, since it's the source of the fall, as perceived by this new left brain. So in the heroic model, as re-invented by the left brain language of ideography, the hero still carries the necessary mythopoetic baggage of supernatural origins, etc. since the origin even of this new hero is a mimetic bioculture type. But he doesn't slay dragons, (for dragons are images of the mythopoetic creation of sound, the dragon being the chaos of touch and heroes, in right brained mimesis replicate the acts of the gods, to remind humans how to re-create the world holographically); THIS hero dismembers himself, his own body. While the sacrifice of right brained mimetic heroes is the detachment from the fruit of their actions, in an effort to subordinate the (left brained) self; this hero renounces the actions of the body; not the fruits of the actions; thereby exalting the virtues of the SELF (the left brain.) And this new hero's ultimate fate? Death of the body! Rejoice! The left brained self can now be liberated from its mortal condemnation: a consciousness with no origin! An immortal self with no taints of materiality!

The only problem for humans who follow this model though are, while we're IN a body, is there any hope for redemption? There are two answers; both yes. Some systematically renounce the pleasures of the flesh by ascetic[10] practices and others transgress the problems of the body entirely by creating a substitute world, invented by the left brain, which can perform all the good functions of the body, but is devoid of the carnal pleasure in the observer, and is therefore not evil. What are these inventions? Ideographic language, which gives birth to the synthetic world of ideas and of technology and theory. And with the invention of this substitute world, the mind/mind shift is complete.

In the next chapter we shall analyze the myth of Oedipus as one of the first vestiges of the new hero of ideography. We shall look at the elements of the mythos itself and compare its elements to those of mythopoetic hero myths in an effort to elucidate the mind/mind split: how it occurred, why it occurred and most importantly, what the consequences of the shift were/are for humanity.

❁ ❁ ❁

Ideographic Creation: The Immaculate Conception

Stephen Hawking, in his book, *A Brief History of Time*, epitomizes the ideographic method of creation in the following quote:

> In order to talk about the nature of the universe and to discuss questions such as whether it has a beginning or an end, you have to be clear about what a scientific theory is.[11]

Why is that true? Because ideographic language is a reflection of the mental process that functions to create it and in this case, in the absence of mythopoetic language, we are left with only the functions of our left brain to ponder these things. When ideographic language takes over, the scientific theory replaces the left brain as the judge, since the origin of the experience IS the left brain itself, the "question poser." Where does the question come from? Curiosity. What brain would have curiosity? The one that didn't know, right? Which brain doesn't know? The one who depends on outside information in order only to agree or veto, the left brain, right? Does the left brain know even the ORIGIN of the question? No. Are you beginning to get the point here? So, since the left brain is perceiving itself to be the originator of the question, it needs to create a substitution system outside itself to analyse it so it (the left brain) can say yes or no. Hence the birth of theory. Is it this theory? Yes or no. Is it that theory? Yes or no. Can we ever be sure? NOT BY THESE CRITERIA. THEREFORE THE LEFT BRAIN INSURES ITS FUTURE. HOW? By creating its own loop: Question, theory answer, question, theory answer... Without the ORIGIN of the body (the ENEMY of the left brain, remember) the left brain can never know it knows, it can only theorize or assume it knows. But deep down somewhere (in the body, in the other brains,) it does know that it doesn't really know, that it is only fooling itself. But since that left brain has become such a powerful internal voice, that has created such a niche for itself in the species by cultures attributing such things to it as biography, salvation, and immortality, it will be a very hard faculty to shut up. Perhaps, on its own it will never shut up.

The myth of Adam and Eve has the solution to this problem built into the myth; actually, it has two solutions. Can we read them anymore? No, because mankind, as the myth says has "fallen."We have become conscious of ourselves and this of course is the FALL. What are the solutions in the myth:

1. The other "tree," the tree of life which WAS NOT forbidden.

2. The vestiges of creation by the criteria of mythopoetic language.

The tree of knowledge of good and evil is the left brain. The tree of life is the right brain, the one that creates paradise and gods and everything else in mythopoetic creation, including the consciousness we count on in order to continue living. By living by the criteria of the fallen ones, though, that is by our left brains, we succumb to the original source of the duality of good and evil: the separateness of ourselves from creation. By eating from the tree of life, we "become as gods" as the story says, because WE create the image, not the demiurge god out there. The message is simple: SALVATION LIES IN THE RIGHT BRAIN TREE OF LIFE NOT IN THE PROSCRIPTIVE VOICE OF THE LEFT BRAIN.

Why? (And this brings us back to the point of the beginning of this chapter,) because the left brain cannot create images, worlds, gods, IT ONLY CREATES SHADOWS, phantoms, concepts without the life force flowing through them. It only asks itself questions so that it can verify to itself, the existence of itself. And THIS SELF[12] is the source of all human suffering. Yet this self, to the exclusion of everything else that is human has become the grand inquisitor of the entire living, universe. Why do we have to know about scientific theory in order to learn about the origin of the universe? Because the only ORIGIN that scientific theory is worried about is the origin of that part of the universe that verifies itself and NO OTHER, the almighty concept maker out there, the left brain. And the origin of that faculty, from any way you want to look at it, is the FALL: when the left brain acts independently from the right brain (and consequentially, the body) and takes itself to be the only god.

CHAPTER THREE

THE HERESY OF OEDIPUS

Introduction

Riddle: Q: What mental act separates humans from animals?

A:

(Is this answer correct? How do you know? What is the answer[1] answered? Have you really solved the riddle? Read on to see . . .)

The heresy of Oedipus is not about incest, nor is it about patricide nor even about murder at all. It is not a polemic about denying divine will, nor is it a tribute to the victory of self. The heresy of Oedipus is not a heresy per se of homeland, family, church, morality, god, country, civilization, or humanity, though the actual heresy produces radical changes in all the above. So what is the heresy of Oedipus?

Riddle: Q: What is the meaning of "Oedipus" in Greek?

A: EITHER, "swollen foot" (from *oidos*, swollen)

OR "he knows" (from *oide*, he knows)

And this fact is the heresy. And here's how and why.

The Development of Writing From Pictogram to Alphabet

PICTOGRAM:	"SUN" AS AN IMAGE	⊙	MAIA
PHONOGRAM:	"SON" AS A HOMONYM OF NAME OF IMAGE	⊙	MYTHOS
IDEOGRAM:	"DAY" AS IT REFERS TO THE TIME WHEN SUN IS OUT	⊙	R. BRAIN MIMETIC
TRANSITIONAL ALPHABETIC:	"S" SOUND AS THE FIRST SOUND IN WORD, SUN	⊙	L. BRAIN MIMETIC
ALPHABETIC:	THE LETTER "S" WITH NO PICTOGRAPHIC CORRESPONDENCE	⊙	LOGOS

Written Language and the Mind/Mind Split

In the last chapter we differentiated between what we have termed heroes of right brained mimesis from those of left brained mimesis by showing how the ideographic interpretation of the former produces the latter. In this chapter we shall take this issue of heroes further to include another type of "myth" of the hero which in reality is no true myth at all, but rather a narrative account of the mind/mind split from left brained mimesis to logos in culture. Specifically we shall use the myth of Oedipus (though there are certainly others) to retrace the actual mental steps, as recounted in the narrative, which brought to light the modes and technologies of left brained dominance in humans.

Before we embark upon the myth of Oedipus it is necessary to say a few things about the nature of the problem of writing, for as we shall see, a good deal of what shall subsequently be called the heresy of Oedipus is in fact a heresy of this human technology. It is necessary to note at the outset that this discussion is not about writing per se; it is about a particular technology that came along with a particular cultural decision among particular groups in reference to one aspect of writing that, unbeknownst to them produced, by repetition, a new hardwire in the brain. And the habitual repetition of this act is what truly altered civilizations forever, for it permanently and biologically changed the significance of what it meant to know.

Consider for a moment the chart at the beginning of this chapter which outlines, by analogy, the development of writing from pictographic representation to alphabetic substitution[2]. As you may recall, the image of the sun is used as the example. In the pictographic mode, the image of the sun is translated as "sun." Thus, any person, if the representation is even remotely similar to the perceived image of the sun in the sky, would be able to recognize that the image represents the physical sun in the sky. This type of abstraction, though representational is actually a mythopoetic mental act as the substitution is an almost tautological image/image transference. In fact, the abstraction is actually a physical representation of the act of mythopoetic creation for the "writer" of the image; since the act of drawing thus is a right brained function utilizing the same mental technological skills as myth making, only altering the medium. As for the "reader" of the image, it is easy to derive a correlation between the two images (the physical sun in the sky and the drawn sun) by the medium of the sense of sight (though the other senses at this point fall by the wayside as far as re- creation is concerned. More about this later.)

The next level of abstraction that takes place in the written language is the use of ideograms. The ideogram in the chart at the beginning of the chapter uses the same image of the sun not to represent the object of the sun, but to represent the idea of "day." This is no longer in the realm of mythopoesis, as the sun picture now serves as a symbol of a concept, day. "Day" is not an image; the sun is an image. And this image now must be logically linked to the concept in order for comprehension to take place in the mind of the reader. Thus the ideogram provides the first representation of the left brain's dominance in the sphere of writing. Still, anyone reading the text would certainly be able to derive meaning between the image of the sun and the concept day without too much difficulty. Obviously, one of the most substantial attributes of "day" is the presence of the sun. This is not, however a mythopoetic derivation, because none of the human senses usually associated with mythopoetic creation, including sight, is used to make the connection between the two. It is the cognitive left brain activity which creates the affiliation first. Then, the VERIFICATION that the connection is correct comes by linking the image of the physical sun, the written sun and then "dayness" together by means of the imagination. So to recap then, the ideographic mode requires the medium of the concept to be understood and only later uses images to verify the conclusion.

Next we come to the linguistic use of the phonogram. Consider the example on the chart. The identical image of the sun that in the pictographic mode represents "sun" and in the ideographic mode represents "day" in the phonographic mode stands for "SON"! Listen, sun, son; they're what we'd call homonyms, words that sound the same but have different spellings and different meanings. So how is one to derive meaning from the phonogram? Is it mythopoetic? Ideographic? The answer is yes yes or no no. Insofar as the original phonogram uses the primacy of sound as its mode of comprehension, it is mythopoetic (in the same way that the pictogram uses sight as the primary sense of comprehension); however, since the similitude of sound doesn't come from the object itself (from either object itself) but from the sound of the NAME of the object, it is a conceptual abstraction and is therefore ideographic in origin. So, in other words, it is really purely neither, though based on a hybrid of both human technologies. At root, though, the understanding of the phonogram is a conceptual one; therefore, the phonogram falls more into ideographic understanding than mythopoetic. But the above notwithstanding, it is still easy to DERIVE meaning from the phonogram using the medium of the human sense of sound. This is a very important point to consider, especially in light of the next linguistic "development."

Pictograms, ideograms and phonograms were all used in early writing systems; in fact most systems used combinations of all of the above in their systems and the reader relied on context in order to derive meaning. As anyone could guess, the labor of writing was enormous until systems of syllabic conventions began to take form, then phonetic systems (where the first sound of the pictogram represented the phonetic sound in the new word), and then finally alphabetic systems. And with the widespread use of alphabetic systems, the heresy of Oedipus was born. To understand the heresy and its unparalleled consequences, consider for a moment the last line on our chart, the alphabetic line. Here, "S-U-N" represents the physical sun in the sky, right? Now, how does one DERIVE this connection? Can you use any of your human senses? Sight? Sound? Taste? Touch? Smell? The answer is a resounding NO!!! Alphabetic substitution is completely devoid of sensation in a way that no other human act is. Its creation then is not a creation through human sensation at all, in fact its "birth" is actually adoptive: no one can derive any meaning from an alphabetic system without someone having told them the rules of the system. And at the point of learning these rules, the human has to learn one more skill: belief. Humans have to rely on the veracity of the system through the medium of the teacher or book in order to truly know the rules. And these rules are totally arbitrary and synthetic; and these rules are not creative in the historical sense (i.e., mythopoetic, based on imagination), and therefore they cannot in any way, shape or form be derived by humans without this mediumship from the outside. And it is impossible to learn all the rules, and one is always left in a quandary, somewhat of an alien in his own skin when it comes to these rules, for they are not part of the internal creative capacities of humans; these skills are outside usurpers. They are the foreigners who come and take over from the outside leaving desolation and destruction of the creative self in their wake. At the root, these rules are born, live and die solely by the mental criteria of the left brain talking to itself in its own conceptual loop, a simple enough origin, but the re-"creation" of this hybrid act wreaks havoc on its victims. In fact, the widespread use of this human technology COMPLETELY ALTERED CIVILIZATION after it was introduced. How? For starters, matriarchy became patriarchy[3]; cults of heroes replaced gods and goddesses; procreative worship became dogmatic worship; words became flesh and hierarchical structures were formed to eradicate the old method of knowledge (by imagining) and instituted in its place the new school of knowledge (which was one that had to be taught and could therefore be controlled). The next chapter examines these and other ramifications in detail; however, it is important to note that even this shift in the brain, from left brain mimesis to logos has its own "myths" and they are the new breed of heroes and of creator gods and divine artisans, created by the eunochs and the priests and the magicians who found in alphabetic writing their own mode of creation, albeit imitative. Remembering that the knowledge of physical creation prior to the development of this system of alphabetic substitution was preserved only for women and the gods, this invention heralded in a new era for male superiority and of course, imperialism of the worst kind: neuro- biological. And the literary remnants of this shift is evident in a certain strand of hero "myths" and creation "myths" with the Oedipus story serving as the model par excellence.

❁ ❁ ❁

Geometry to Gematria and the New Masculine Knowledge: The Logos

It is necessary here to take a brief diversion in order to outline how it is possible that a culture may be able to completely turn around simply as a result of the introduction into humanity of this logos loop. The very obvious path is one that is fraught with absurd, even comical inferences; however, it is clear that it was in fact through this path that the mind/mind split changed civilization forever.

What is mythopoetic knowledge? It is knowledge of creation; the creation of humans, of forms, of the gods, of the cycles of nature, of the phases of the moon, of the times for planting, of the times for harvest, of the gestation period to bring forth another human from the womb of the abyss. Because of the special relationship that women have always had to the knowledge of the cycles of creation, the repository of knowledge, in mythopoetic cultures was within the company of women. From the first recognition that the phases of the moon corresponded to the monthly cycles of women's bodies, women began the art of calculation.[4] Linguistic evidence of the connections between motherhood (the act of creating another human form) and measurement (the knowledge of the duration and geometric proportion necessary to create) is substantial. For example, Sanskrit, *matra* means mother, but it also means measurement. Greek, *meter* likewise means both motherhood and measurement. Consider also the following English words, based on these roots, which all have to do with this mythopoetic knowledge: mathematics, geometry, mensuration (and countless others; forgive the pun). It is interesting to note also that the root word for moon in Indo-European (*manas*, *mana*, *men*), was also the root for mind; thus creating a sort of mythopoetic loop of its own with motherhood, measurement, the moon and mind all being related functions, according to linguistic criteria.

The ancient goddesses of mythopoetic maia biocultures were venerated not only for their ability to create human form, but also for their capacity, due to their knowledge of measurement, to create all the civilized arts, such as agriculture, weaving, building, writing (mythopoetic), calendars, and of course mathematics and music.[5] The maia bioculture then saw the One as feminine, because OUT OF HER, all of human creation is born: including the creation of humans themselves, and civilization. This deity was known to the people as the triple goddess: first, because of the triple role of the sensations of touch, sound and sight in the formation of the material world by the criteria of mythopoetic creation. Later, these functions were embodied within an archetypal imaginative structure and she was personified as the goddess of creation, preservation and dissolution, for just as the cosmic cycles of the moon waxed and waned, constantly altering form, though eventually returning to its original dark, formlessness, so did the life-giving forces of the goddess/woman. And so it was she who was the original Trinity, representing the three phases of feminine life: the virgin, the mother and the crone. (Note that the masculine role in creation was not causally linked to the capacity to make form in this bioculture.)

Mythos biocultures venerated the holy couple as the source of knowledge, with the feminine role retaining its earlier position and with the masculine role being the first "creation" of her. Hence, the male links the two worlds of existence before manifestation and the first manifested form, so he, in a way becomes the "father" of things in material form, since he is in fact the prototype. The fusion then, through a *hieros gamos* (sacred union) between the female and male is what constitutes mythos knowledge, for it is through the harmonia of the opposites that all creation is repeatable. In this bioculture, the triple goddess is personified as different deities, with each one representing a single aspect of the creative process; likewise is the masculine role of creation represented, with the triple god roles being: the sky father, whose function it is to penetrate the female; the god of death, who represents the phallus inside the womb/tomb; and finally, the god of the "masculine waters" that cause creation to be brought to fruition. And when the sky god mates with the earth goddess, and they give birth to a child; that threesome then becomes the mythos trinity: father, son and child. Since this is the bioculture that emphasizes the dual role of creative powers (masculine and feminine), it is correspondingly the bioculture where music and geometry become linked (the former as the feminine origin of the latter), since, as Goethe put it, "Geometry is frozen music."

As we said earlier, the male is the prototype of created form, so in right brain mimetic biocultures, which focus on form, knowledge IS knowledge of form-making: and the demiurge now takes on a role a sole creator because the feminine force is not in material form, merely in geometric or musical form (in other words, it doesn't have three dimensionality.) It becomes easy here, for the demiurge (read here, masculine) forces to appropriate knowledge. But knowledge here is the knowledge of the power of form-making, which could include everything that goddess knowledge provided AND the assertion of the superiority of the masculine force in the *hieros gamos*, because, after all, THE MASCULINE COMPONENT BROUGHT WITH IT, FORM. In right brain mimesis, the demiurge is primarily though a god of form making in the material sense: an artist, a craftsman-one who uses his triple powers of creation (sight, sound, touch) to invent objects in the material world. He is also the prototype of the magician, whose knowledge of this kind of creation yields him much power amongst other men. It is necessary to note, however that this bioculture, though masculine oriented, does not eliminate the feminine role in creation; it merely creates a new substitution system for the creation of material, albeit non-human form; one that males alone can create as easily as females can. Note too that in right brain mimesis certain female characters also fall under this bioculture; for example wisdom goddesses such as the Greek Athena, whose primary function is not procreation but material or cerebral creation.

All this changes with left brain mimetic biocultures. In the left brain mimetic biocultures, the demiurge is the SOLE CREATOR OF ALL THINGS, ALL KNOWLEDGE, ALL LIFE, ALONE. With the appropriation of the civilized arts from the hands of women and the capacity to re-create them at will, the demiurge also becomes an imitation of the maia goddess: the One, envisioned as the Trinity. Here, there appears a simple substitution system of the functions of the maia goddess directly to the left brain mimetic god. The first aspect of the masculine trinity is the father; the second is the son, namely, his actual son and the third aspect is the semen which provides the continuity of form-making of humans (preferably masculine humans), in the future. In short, the mysterious trinity of the left brain mimetic bioculture is none other than the personification of the masculine role in procreation, WITHOUT the appearance anywhere of the feminine role. In fact, in this model, women are considered "virgin" simply because THEY contain NO element of the creative force within them, according to this model. Their only function is to be the repository for the "god to come."In short, the god of left brain mimesis BECOMES the trinity of the triple goddess; he becomes the creator of everything and everything eventually returns to him (How? Don't ask. o.k. The answer is that he creates by masturbation and everything is returned to him by his eating it.) (See the myth of Ra in the appendix and the myth of Kronos in the appendix, for the details.)

What are the signs of this god's sovereignty? The tablets of the Laws he creates, which are imbued with magical meaning that ONLY the elect (the priesthood) could comprehend. And this magical meaning, of course is none other than the NUMERICAL PROPORTIONS OF CREATION STOLEN FROM THE MAIA GODDESSES, but RE-CREATED by this new male god, into the form of GEMATRIA.[6] (Gematria is the practice of interpreting words and phrases by their numerical values. Remember that in ancient writing systems such as Babylonian, Hebrew and Greek, numbers were represented by alphabetic symbols, thus a letter could have two meanings; the exoteric letter meaning OR the esoteric, magical number meaning.

As a result of gematria, new pantheons were set up, proper spelling became significant and there was created a huge gap between the knowers of knowledge and the general public. Of course, the knowledge didn't change, it was only hidden in a new form so that only the "elect" were privy to its contents. It is also interesting to note that, while gematria was in fact originated OUT OF geometry, very quickly the original geometric forms all but fell by the wayside for the introduction of this new hocus pocus study of the "words made flesh." This is the origin of magic in religion.

Finally, we come to the birth of alphabetic language. With the shift to alphabetic language, gematria completely took over as the knowledge of God. In fact, the WORD BECAME GOD.[7] For after all if the geometric form of creation is in this word, then this word IS the geometric form of creation, right? Well, no, but, believe it or not THIS logical fallacy is the foundation of the logos bioculture's form of knowledge. As above, so below became, as below (since we know the secrets and possess them here in these words), so above (there is nothing more to creation, knowledge, god, immortality THAN THESE WORDS; therefore, WE KNOW THE PLAN OF GOD. LITERALLY. No pun intended here, either.) And this is the context in which we find ourselves in the myth cycles of the logos heroes. And the deity of the logos bioculture? The trinity of time: of the past, present and future; the invariants that survive both life and death and are totally outside of the bonds of mortality; in other words, the theory of time has itself become a god.

THE BIOCULTURAL TRINITIES AND THEIR MEANINGS

The Maia Trinity: The Triple Goddess

As the CREATOR, PRESERVER, AND DESTROYER, the triple goddess in the religious systems of the maia bioculture represent the three aspects of creation as exemplified by the three roles of woman. The creator aspect refers to the virgin goddess who gives birth to the child; the preserver aspect refers to the role of woman as the repository for the skills necessary to maintain a harmonious society as well as the obvious maternal aspects of nurturing and the like. Finally, the destroyer aspect represents the crone; the woman past the child bearing age whose creative "wisdom" is maintained within herself. It is she who represents the return to chaos; that is, the return to the unity of the beginning and also the capacity for re-creation now that all has been returned to the origins. A typical example of this triple goddess is the Indian triple Kali, whose white, red and black manifestations represent the trifold aspect of the life of a woman and correspondingly, of the cosmos at large.

Other examples include:

-Egyptian:	Isis
-Etruscan:	Uni
-Druidic:	Diana Triformis

The Mythos Trinity: The Dual Roles of Procreation

As the mythos bioculture realizes the HIEROS GAMOS as the source of all creation, the trinities of this bioculture likewise represent the triple feminine role, as outlined above as well as the triple masculine role, as described as follows:

> The impregnator (the sky father god, the thunder wielder, etc.) is the masculine force that enters the woman; the god of death (the god "entombed" in the underworld/womb) is the act of procreation (penetration) itself; while the god of the "masculine waters" is the "masculine deposit" required for procreation. Thus, for example in the Greek pantheon, Zeus, Hades and Poseidon represent the three aspects of masculine creation in that Pantheon: Zeus is the sky god; Hades is the lord of the underworld and Poseidon is the "masculine waters."

In this bioculture the trinity is reflected in the father, mother and child motifs. The best example of this type of trinity is the models in Greek myth of the voluminous examples of Zeus' relationships with goddesses and mortal women which yielded the population of the younger generation of gods, goddesses and mortal heroes. For example, Zeus and Maia produced the god Hermes; Zeus and Alcmene produced Heracles; Zeus and Hera produced Hebe, etc.

Other examples include:

-Egyptian: Isis, Osiris, Horus
-India: Maya, Ganesha, Buddha
-Babylon: Tiamat, Kingu, Marduk

The Right Brain Mimetic Trinity: the Demiurge

In right brain mimesis, the triple role of the sensations of sight, sound and touch form the basis of the trinitarian aspect of demiurge, for as master of the elements of creation, he is seen to have dominion over the animal, mineral and vegetable kingdoms, for by knowing the how of creation, he is able to mutate and thus transform form, and this is his primary role. The masculine forces, as described above, are likewise considered in the process of right brain mimetic creation, and they are all ascribed to a single masculine deity: a demiurge, who, by his power of control over these three aspects of masculine creation, and the three levels of material form, is able to sublimate (to use a Freudian term) his procreative tendencies and apply them to other aspects of creation, such as the creation of art, crafts, architecture, literature, etc. In other words, the world of material form, but not human form. Any demiurge god of any Pantheon, for example the Greek Hephaestus, represents this trinitarian god.

Other examples include:

-India:	the Rhebus,Tvastra
-Egyptian Memphis:	Ptah
-Greece:	Prometheus, Dedaelus

The Left Brain Mimetic Trinity: Father, Son and Holy Spirit

In left brain mimesis, the functions or powers ascribed to the masculine trinity now become personified as physical realities: the sky god becomes the FATHER; the dead god becomes the SON who must be "sacrificed" so that the future generations are "saved"; and the HOLY SPIRIT is none other than the "life giving" masculine waters of the mythos bioculture, now transformed into a great mystery, simply because of the biocultural shift in substitution systems! This is the bioculture that heralds masculine forces as the supreme and sole forces of "God," because that's exactly what they are: three aspects of masculine creation transformed into a theoretical substitution system by the left brain. The Christian Trinity is a prototype of left brain mimetic trinities.

Other examples include:

-Phoenician:	El, Baal, Yamm
-Egyptian:	Ra, Shu, Tefnu
-Indian:	Brahma, Visnu, Shiva

Logos Trinity: Past, Present and Future

The logos trinity is simply the PAST, PRESENT and FUTURE in historical time. There is no correlation between human life and the vacuum of history which precedes us, lives with us and undoubtedly continues long after we are gone.

Let us return now to the myth cycle to delineate in mythic terms, how this horrendous shift took place through the house of Oedipus, and with what results for all of us.

❁ ❁ ❁

Words Made Flesh: The Logos "Myth" Cycles

The paradigm of the logos hero retells the story of the development of the human faculty of theorizing; the "creation" of the imaginatively sterile left brain. Since there are no authentic images in this creation, these stories are really "rational" explanations for the occurrences of certain events in time, sort of proto-history. Nevertheless, the paradigm of these stories is discernable and it is as follows:

1. The hero is born to mortal parents who had received grave portents from an oracle regarding the birth of the child.
2. The child is locked up in a chest and thrown into the sea to be drowned and/or is reared inside a cave.
3. The child is raised by foster parents.
4. The child has either a deformed leg, foot, knee or thigh OR is born out of a male/god's thigh.
5. He demonstrates superior talents. They are EITHER cleverness of wit (solving oracular statements, teaching in riddles or parables) OR demonstrated accomplishments with their hands (craftsmen, metalworking, etc.)
6. These talents land him much notoriety as the true savior of his original fatherland, though his presence in his homeland causes unparalled strife

there.

7. He either has no children or his children do not enjoy a positive fate.
8. As a ruler he breaks old laws and prescribes new ones.
9. He takes on the role of the sun-god, while any feminine counterpart in the myth is labelled by a lunar appellation.
10. He dies on a hill and is apotheosized as a divinity.

This is the basic pattern of the hero of the word.[8] His actual claim to fame is that he is the embodiment of the technology of artificial creation, or transmutation; that is, the art of taking one element that already exists in form and altering it to make a new form. This is the earliest cultural attempt at male creation, and it is the kind of creation that mimics the functions of the left brain when it acts in the capacity of its own interpreter of its own functions. Thus, as a literary explication of male creation, naturally, these myths must recount a dual origin of these men; one through the instrumentation of the female but the other "more important one" through the re-birth of the male into his new, albeit adoptive home. Every logos hero of necessity then must be:

1. Born under the auspices of a fatal PROPHECY. Prophecy is crucial to these hero stories because it sets up at the outset the criteria under which the hero emerges: when a statement is made devoid of context in an almost apocryphal mode (from the perspective of the mythopoetic mind), and necessitating interpretation by the receiver of the information even though the one receiving the information has no means by which to interpret thus. This uneasiness on the part of the receiver of the prophecy begins the entire chain of events that lead to

the ultimate "birth" of the new hero.

2. Twice born with only the second one counting : hence the part of the myth where the child is thrown into the sea or reared in a cave. (These are the second births which are SYMBOLS of female wombs without the females.)

3. Reared by foster parents because his true origin is both unknown and unimportant. (As when the left brain, unaware that the images it perceives as totalities are in fact manufactured by the right brain, uses these images to create new realities:concepts.)

4. Deformed or otherwise affected by his leg, foot, knee (read here *membrum virile*). This is important because his form of creation must be distinguished from PROCREATION, of which this hero is not a part. In an alternate form of this myth, the hero is actually born out of the "thigh" (you know what to read here) of his father.

5. Known for his superior talents of either solving riddles (a natural talent for the left brain) or creating ingenious technical objects (in imitation of the demiurge gods, whom they usually venerate). Again, this mode of creation is not imaginitive, but cognitive.

6. Viewed as great saviors by some in their homeland, but really mistrusted by others. This is a natural conflict that anyone goes through when their head tells them one thing but their heart tells them something else, only in the narratives, the conflict becomes more global.

7. Childless, because this is not their goal of creation. If these heroes do

have children, it is usually viewed in these myths as the primary source of their downfall.

8. A prescriber of new laws, since the old modes of knowing become obsolete. The laws then, replace the internal capacities for knowledge from the earlier mythopoetic bioculture and the hero, since he imposes the laws, becomes the most powerful member of the group FOR HIS NEW KNOWLEDGE.

9. A physical symbol of the sun god on earth, since the sun god, in mythopoetic biocultures is the closest thing to a logos creator. The sun, as the provider of light (insight, illumination, etc.) becomes the SYMBOLIC associate of the logos hero, since it gives humans the capacity to "see" the world of forms; an act mirrored by the functions of the logos hero/left brain when it creates its own conceptual realities, thereby acting by analogy, as the sun god within its respective mode. Enter the concept, "as above, so below." The moon, as the image of mythopoetic creation, is relegated into a subservient position in these myth cycles, which after all, are meant to exalt the new knowledge, not the old one.

10. Raised up at his death, though no one is quite sure how, why, where or what for. The answers to these questions, according to the myth cycle are intentionally left a mystery; they are to be believed on faith, for after all, it is impossible to DERIVE any conceivable meaning out of this event. And as we know, we are not supposed to, are we? After all, could you, reader ever be able to understand the profound creative

mystery inherent in the relationship between S-U-N and the image of that blazing fireball way up there in the heavens? The answer is of course not, because THERE IS NO PROFOUND SECRET. The myth of the secret is the heresy of Oedipus and this heresy has to do with the conspiracy of this kind of knowledge: sterile, impotent, evil, degrading, de-humanizing, unintelligible dogma. And this IS the heresy of Oedipus. A brief retelling of the mythos displays the blatant references of this mind/mind shift in the bioculture of ancient Thebes and also sets up the foundation for how and why this shift gained control of our minds forever.

✿ ✿ ✿

The Mythic Ancestors of Oedipus

Zeus and Io[9]

The usual origin of the house of Thebes is traced back to the myth of Zeus' abduction of Io, a virgin priestess of the goddess Hera from Argos. According to the myth, Io received an odd dream, telling her to go to the meadows of Lerna and yield to heavenly love. When the troubled child told the dream to her father, he immediately consulted the oracles and was told to banish the girl from his homeland, which he did. Meanwhile, Hera, Zeus' wife who was jealous of Io, changed her into a white cow, whereupon Zeus, taking the form of a bull, had his way with her. Hera, aware of these events, demanded that Zeus produce the cow to her as a gift of fidelity, which he did. Hera then ordered the monster, Argus panoptes (many-eyed) to guard the cow lest Zeus not take her away. Zeus, who of course wanted Io back, ordered the god Hermes to steal Io away. Hermes managed to lull Argus to sleep by telling him boring stories and playing the shepherd's pipe, and then, once the monster was asleep, he chopped off his head. This was not looked upon highly by the other gods who stoned Hermes for having done this act.

Unfortunately, Io's troubles had not abated though, for Hera immediately seized upon a new way to torment the poor girl/cow. The ghost of Argus was turned into a gadfly, who continuously stung the poor Io, as she aimlessly wandered about the world, slowly regaining her human shape. She traveled through the Bosporus (cow's passage), finally coming upon the Nile River, where exhausted, she rested. Zeus, seeing the girl/cow, gently touched her and returned her to her former shape, and there she gave birth to a son, Epaphus. Later Io was worshipped in Egypt as the goddess, Isis.

Epaphus and his children

When Epaphus was just newborn, Hera ordered the Curetes to steal him away, which they did. (The Curetes , from kouros, gr. "male youth," were a band of male diviners whose method of divining was to dance about as a group clashing spears and banging drums.) Io roamed all over the world in search for him, and passing through Syria, she learned that he was being reared by the queen of Byblos, in Phoenicia. When Epaphus was returned to Io, he became the king of Egypt and married Memphis, the daughter of the Nile god. Epaphus named his settlement after her (Egyptian Memphis) and they had two daughters: Libya and Lysianassa. Both of these girls fell prey to the god, Poseidon, and they each had children by him. Libya gave birth to twin sons, Agenor and Belus. Lysianassa gave birth to one son, Busiris. Busiris remained in Egypt and ruled there as king and was known for his odd custom of sacrificing foreigners to Zeus. It was said that this custom began after Egypt was plagued by a nine year famine. An oracle from Cyprus told the king that annual sacrifices (human) must be made to insure the continuity of his land's fertility. Unfortunately, the seer was Busiris' first victim. And even more unfortunately, Heracles was another intended victim. When that hero was about to be sacrificed, he sprang free of his chains and instead sent that king and his son to their deaths. Thus ends the line of Busiris.

Belus and the settlement of Egypt

The land belonging to Libya was divided between her twin sons, Belus and Agenor. Belus remained in Egypt and married the Egyptian, Anchinoe, a daughter of the Nile god. Belus and Anchinoe had a daughter, Thronia; twin sons, Aegyptus and Danaus; and some say, another daughter, Dido. He was also reputed to be the ancestor to the Persian kings. Thronia married Hermaon and gave birth to a son, Arabus after whom Arabia was named. Dido founded ancient Carthage, specifically, Byrsa, the most famous city there. Aegyptus remained as king of Egypt; in fact, Egypt bears his name. And Danaus was forced to flee to Argos, original homeland of Io, to claim the throne there, because his twin did not wish to share the kingdom and was plotting to kill him.

The Myth of the Danaides (daughters of Danaus)

Both Aegyptus and Danaus had many wives, who gave birth to many children. Aegyptus had fifty sons; Danaus, fifty daughters. In an effort to maintain the entire kingship, Aegyptus plotted to marry off his sons to their cousins and then to kill his brother. Danaus, remembering an oracle that said he would be killed by one of his nephews, took his daughters and went to Argos to claim kingship there, as he was in fact a true descendant of Io. Aside from the oracular warning, the daughters were horrified by the marriage proposal because of the semi-incestuous nature of it and also because they were repelled by men in general. With the aid of Athena, who taught them how to construct a two prowed ship, the group landed in Argos.

Argos was in the state of a terrible drought, because Poseidon cursed the land when its inhabitants chose to worship Hera instead of him as their patron. When Danaus needed water in order to perform a religious rite, he sent all his daughters out to look for a spring. One of them, Amymome, while searching, came upon a stag and began to chase it down. She threw a spear at the stag, but missed, bruising a satyr instead. The satyr began to pursue her when Poseidon saved her. With soothing words, that god then wooed her into his embrace and she begot a son by him (Nauplius, the great seafarer). In exchange for her acquiescence, Poseidon struck his trident into the ground and created a sacred spring for Argos. Because of this, Danaus and his family was welcomed into Argos. The then king of Argos, another rightful descendant of Io, promised to defend them against the wrath of even Aegyptus, but Danaus wanted the throne. The two gave speeches to the people of Argos, each arguing why they and not the other should rule. Although at first the Argive people chose their own king, an ominous omen made them change their mind. That night, while all were asleep, the chief bull of the royal house was slain by a wolf. The people interpreted this to mean that Danaus, the outsider, should rule, and they turned over the throne to him. Upon his ascension to the throne, he dedicated a temple to the god Apollo in his wolf incarnation. Meanwhile, Aegypus and his sons were planning to siege Argos, out of fear that Danaus would attempt to overtaken them. The invaders won and were given the daughters in marriage; however, Danaus had plotted a deadly end for his nephews. Each daughter was armed with a dagger and they were ordered to "behead" their bridegrooms on their wedding night. Forty-nine of them did, and afterward threw the "heads" of their husbands into the sea. One daughter though, Hypermnestra refused to kill her groom, Lynceus, because he had respected her virginity, so she fell in love with him and aided him in his escape from the city. When Lynceus reached safety, he flashed a light beacon to his wife to let her know.

Hypermnestra was tried by her father for disobedience when he learned of his nephew/son-in-law's escape, but the people of Argos set her free. Then, later, Lynceus returned to Argos and seized the city, putting Danaus on trial for murder. Both the Egyptian and the Argive people tried the case and found Danaus guilty. He was put to death and his remaining daughters were each awarded to men as booty for winning footraces. Then Lynceus and Hypermnestra took over the rule of Argos. They had a son, Abas who ruled after his parents. He married Aglaea, and had a daughter, Idomene, and twin sons, Acrisius and Proetus.

Acrisius of Argos

King Abas desired that his sons jointly rule Argos; however, because of their constant fighting, and also because Acrisius accused his brother of seducing his daughter, young Danae and fathering a son by her, Proetus was forced into exile. He went to Lycia and there married the king's daughter, Stheneboea. The king of Lycia helped Proetus by sending an army to Argos in an attempt to overtake the city, but the battle came to a draw between the two brothers, causing the kingdom to be divided between them. As a result of the division, Acrisius maintained the seat of the kingship in Argos and his brother moved to Tiryns and ruled there. Acrisius married Euridice, of Sparta and they had a single child; a daughter whom they named Danae. Now Acrisius longed to have a son, so he went to seek the oracle's advice on the matter, but rather than answering his question, the oracular message was that should Danae have a son, that child would kill him. Returning home, and determined not to suffer the fate of the oracle, the king imprisoned his daughter in an underground chamber made of brass. Zeus, who became enamored with the girl, came to her in a shower of gold, and she gave birth to a son whom she named Perseus. The king was outraged and as a result, locked both the mother and child in a large chest and threw them into the sea. Perseus, needless to say grew up and became a famous hero (see below). And the oracle came true; he killed his grandfather by accidentally hurling a discus at him during some athletic competition. And Danae emigrated to Italy and there founded Ardea, a coastal city south of Rome. Perseus was ashamed to inherit Argos from his grandfather since he killed him, so he traded kingdoms with his cousin, and ruled Tiryns instead. Perseus married Andromeda and had six sons and one daughter. Among the most notable members of his descendants are Perses, his son, who became king of the Ethiopian court and was the ancestor of the Persian kings (Persia being named after him) and Heracles, his great grandson (actually, step-greatgrandson or half "brother," as Heracles' father was also purported to be Zeus).

Proetus of Tiryns

Tiryns, the city of Proetus, became known for its gigantic fortification walls, which the cyclopes themselves built for the king. He and his queen, Stheneboea, had four children: Megapenthes, a son; and three daughters named Lysippe, Iphinoe and Iphianassa. Despite the cyclopean fortress, the fate of this city became threatened by a series of auspicious events.

The first event occurred when the queen fell in love with young Bellerophon (see below) who had been sent to the court of King Proetus. Though he spurned her advances, the queen told her husband that the young man tried to seduce her, whereupon the king sent Bellerophon to Lycia, his father-in-law's kingdom, with a note asking the king there to kill him. But Bellerophon didn't die there; instead he married Philonoe, queen Sthenoboea's sister. When Sthenoeboea heard about the marriage she killed herself.

The daughters of Proetus also suffered a dreadful fate: they and all the other Argive women were driven mad, believing themselves to be cows. It seems that the women were driven mad, either by the god Dionysos, because the women refused to honor his feasts, or else by Hera, because the girls had somehow defiled her sacred temple. Nevertheless, the king was forced to give up a third of his kingdom to the seer, Melamphus, in exchange for curing the girls. Melamphus, who learned the rites of Dionysos from king Cadmus (see below) was able to restore the girls to their former sanity, but not before extorting another third of Proetus' kingdom for his brother, Bias. As for the daughters' fate; there was one fatality, Iphinoe. As for the brothers, they each married one of the king's other daughters and had children by them.

The fate of Proetus' only son, Megapenthes was equally troublesome. As we stated before, he traded his kingdom of Tiryns with Perseus for that of Argos, where he reigned as king. According to some accounts, he eventually killed his cousin Perseus, because Perseus had turned Megapenthes' father to stone by showing him the Gorgon's head. In turn, Megapenthes was killed by Abas (see above) to avenge his father, Lynceus. The kingship of Argos then fell to his son Anaxagoras, then his son, Alector and then his son, Iphis. When both his son and son-in law died, Iphis left the throne to his brother's son.

✿ ✿ ✿

The Heritage of Oedipus from the House of Cadmus

King Agenor of Tyre[10] had a daughter, Europa and four sons, Phoenix, Cilix, Thasus and Cadmus. One day, Europa was playing by a meadow when Zeus appeared to her in the form of a beautiful white bull and whisked her off to the island of Crete. There she gave birth to a child by Zeus named Minos. Through her lineage, the great Minoan civilization was born.[11] All of this was unbeknownst to her father who, frantic to learn of his daughter's whereabouts, requested each of his sons to go out and search for her, telling them that they should never return without bringing her back. And none of them ever did.

Phoenix settled in the coastal area of the Near East, north of present day Palestine and became the ancestor of the Phoenicians. Cilix settled in Cilicia (present Turkey) and founded a civilization there. Thasus landed on an island in the north Aegean, a land that still bears his name, and settled there. And Cadmus settled in an area in southern Boeotia in Greece, and founded Cadmeia, later called Thebes.

The circumstances of Cadmus' founding of Thebes are as follows:

When Cadmus left Tyre, he traveled to Greece through Thrace, bringing with him from his homeland the knowledge of alphabetic writing. Upon arriving in Greece, he went to the oracle of Apollo at Delphi to ascertain where he could find his sister. The response from the oracle was to give up his search for his sister, and rather to follow the first living thing he saw on leaving the temple. He was to wait until that thing rested and then sacrifice the beast, and upon that site, found a city.

The first living thing that Cadmus saw was a beautiful cow with an unusual white moon shaped marking on its side. Cadmus and his companions did as he was told and they followed the cow. When the cow came to rest, Cadmus asked his men to fetch some water so that they could properly sacrifice the beast. When his companions failed to return after a long while Cadmus went out searching for them. He came upon a spring and discovered that all of his men were killed by a dragon who guarded that spring. The dragon, unbeknownst to Cadmus, was sprung by the war god, Ares. Cadmus killed the dragon, but was now left totally alone to found his city. Athena, who was fond of Cadmus, told him to take the dragon's teeth and plant them, which he did. Out of these teeth grew a race of men, Spartoi (Gr. "sown men"). Cadmus threw stones among them and they, thinking they were being attacked by the others, fought each other to the death, killing all but five.

These five "men" plus Cadmus then founded the city after Cadmus served an eight year sentence to Ares for killing the dragon.

Cadmus was given Harmonia as a wife. Harmonia was the illegitimate child of Aphrodite and Ares (for Aphrodite was married to Hephaestus, the lame god, at the time of Harmonia's birth). Hephaestus gave the couple a robe dyed with ills and a necklace full of curses as a wedding present. Although the gifts looked spectacular, they were intended to bring disaster to the couple and all their descendants.

Cadmus and Harmonia had four daughters and one son. Each of the daughters and their families came to unhappy ends. Ino had two children, both of whom were killed (one by her the other by her husband) when Hera drove the couple mad. Autonoe had a son, Actaeon, who was dismembered by his own hounds while out hunting one day. This happened because he came upon the virgin goddess Artemis while she was bathing naked, and she was so infuriated that she changed him into a stag and his own hounds killed him. Semele had an affair with Zeus and was consumed by lightning when she demanded to see him in all his glory. They had a child, Dionysos, the Greek dying/rising god of vegetation, wine, sexual frenzy and drama. Agave's son, Pentheus briefly took over the throne of Cadmus, but he too was murdered by his mother while she was intoxicated by the spell of a Dionysian rite. Thus, all of the daughters of Cadmus and Harmonia fell by the curse.

Polydorus was the only son of Cadmus and Harmonia. He married Nycteis, daughter of one of the Spartoi and had a son, Labdacus. Unfortunately, both Polydorus and Labdacus died young, leaving the kingdom of Thebes in the hands of outsiders until Laius, the young son of Labdacus, became of age and was able to reclaim the throne for his lineage. Thus, Laius the child was sent to king Pelops for refuge, until he was of age.

While at the court of Pelops, Laius became enamored with Chrysippus, the young son of the king. This was the first case of such a relationship in all of Greece. As a result of this passion, Laius kidnapped the child; fortunately Pelops recovered him, but not without first putting a curse on Laius and his entire lineage. Hera, goddess of marriage was likewise aghast at this transgression and caused the oracle at Delphi to reinforce the curse of Pelops by stating that if Laius should have a son, that son would kill him.

Laius returned to Thebes, took the throne and married Jocasta, daughter of another of the Spartoi. Laius, disinterested in Jocasta, and fearful of the oracle, resisted laying with his wife until one day she intoxicated him and she conceived and bore him a son. Fearful, Laius ordered that the child be exposed in a meadow sacred to Hera, that the goddess might be appeased with the child as sacrificial offering. Instead, a Corinthian shepherd, unable to kill the child whose feet had been pierced in order that he not move, put the baby in a chest and set it adrift out to sea. The queen of Corinth, a childless woman, was out at the seashore washing her clothes when she came across the child in the basket. She took the child to the palace and reared it as her own, naming him Oedipus, "swollen foot."

❁ ❁ ❁

Oedipus and his Fate

Oedipus grew up as the heir to the crown of Corinth, believing that the king and queen were indeed his true parents, until one day, a drunken companion revealed to Oedipus that he was not the son of his "father." Oedipus, upon hearing this accusation went straight to Delphi, to ask the oracle there about his origins. The oracular response was troublesome; she told him nothing of what he asked, but rather she predicted that he would kill his father, marry his own mother, have children by his mother and become an offense to all mankind. The horrified Oedipus left the sanctuary running, until he came to a place in the road known as the Cleft Way. Here the roads to Delphi, Corinth and Thebes merged. Oedipus, in an attempt to avoid the Delphic prophecy, turned left instead of right and began his sojourn to Thebes. Moments later, he came upon a chariot carrying an old man and three attendants who attempted to force Oedipus off the road. Oedipus, known for his tempestuous nature, refused to yield. The man in the chariot smashed Oedipus in the head with the two teeth of the charioteer's goad, at which time Oedipus killed him and two of the three attendants. One of those men was Laius, the real father of Oedipus, who was on his way to Delphi to ask for help against the Sphinx, a monster with a lion's body, a snake's tail, bird's wings and a woman's head, who was ravaging his city. Laius had been told by the local seer, Tiresias, that the reason that the Sphinx was killing people was because of Laius' passion for Chrysippus, but Laius refused to believe the seer, choosing instead to take the trip to Delphi and seek Apollo's advice. Some say that he had an awful dream connecting his "dead" son with the riddle of the Sphinx.

The riddle we are speaking of here is the ingenious method by which the Sphinx destroyed her victims. All those who could not solve the riddle were destroyed. As you may recall, Oedipus was on his way to Thebes when he too came upon the Sphinx, who was guarding the gates to the city, terrorizing the town with her riddle. Upon seeing Oedipus, the Sphinx posed her question to him:

> "Something on earth is two-foot, three-foot, One voiced. Nothing on earth, so variable, of all that creep or swim or fly the air. And when the number of its feet is most, then least of all the swiftness of its gait."

Oedipus responded:

> "Listen, wicked winged Muse of death, Hear my voice, though little to your liking. The thing you sing is man, a creeping creature, four-footed from the cradle, then on two feet, till old age bends his features to the earth, his third foot being his supporting staff."

With this response, the Sphinx threw herself off the cliff and she was dismembered. Oedipus was welcomed into the city as its savior and married the newly widowed queen, his mother.

Oedipus the King

Oedipus ruled uneventfully for several years. He and his "wife" gave birth to two sons and two daughters. Then, a terrible plague broke out in Thebes. Women became barren, crops failed and the entire town was on the throws of utter stagnation. The people asked Oedipus, the wise one, why this was happening. He sent his brother in law, Creon to Delphi to find out. The reason: Laius' murder was not avenged and the murderer was polluting the land. Oedipus then pronounced the following decree:

-That no one should shelter the slayer of Laius

-That a reward would be offered for any information regarding the slayer of Laius

-That he himself would lay a curse on the slayer, and pledged to track him down, as if Laius had been his own father.

The seer Tiresias was goaded into coming into the court to aid Oedipus in finding the slayer. The seer, after tolerating much jeering from Oedipus, finally named Oedipus himself as the killer. Oedipus accused the seer of conspiring with Oedipus' brother inlaw to usurp the throne away from him which further infuriated the seer until finally Tiresias proclaimed," You (Oedipus) have offended against your own people both dead and living. The double curse of father and mother shall drive you from this land. Darkness shall cover your eyes. You shall learn what you are and who those are that look on you as father."

Oedipus then denounced Creon, for setting the seer up to say these things. Creon denied that he had anything to do with it. Jocasta, his wife, trying to ease Oedipus' mind, told him not to believe anything that seers say, for she knew for a fact that many oracles never come true; like the one that said that Laius would be killed by his own son. He was killed by robbers, she told Oedipus, at the place where the three roads met . . . This fact troubled Oedipus, when he recalled the old man in the chariot whom he had killed there and he forced the issue and found the escaped attendant who had survived the slayings there. Then, a message came from Corinth of the death of the king there, the "father" of Oedipus. This fact confirmed to Oedipus the untrustworthiness of oracles, as he knew it was not he who killed him; nevertheless, he refused to return to Corinth to become king lest he fulfill the second part of the prophecy that he marry his mother. The Corinthian messenger, in an effort to calm the worries of Oedipus told him not to worry about THAT prophecy as he was not even the true son of the queen! Oedipus continued unraveling the truth when the herdsman who was supposed to have exposed the child of Laius at birth confessed that he had given the boy to a Corinthian shepherd instead. One thing lead to another and Oedipus learned that indeed he had already fulfilled the dreadful prophecy of Delphi.

Jocasta hanged herself; Oedipus blinded himself by gouging out his eyes with her brooch, and falling under his own curse of banishment, left Thebes as a blind beggar, with his daughter, Antigone, as his guide.

The End of Oedipus

Oedipus wandered as a blind beggar and his two sons fought over the throne, finally causing Oedipus to curse even them, that they should die at each other's hand (which they did). One day the old and weary Oedipus landed in Colonus, outside of Athens, at a grove of the Furies. Though the townspeople urged him to leave, he knew he'd be safe there, for an oracle had told him that there he would be relieved of his suffering. King Theseus was sent for by the people in an effort to get rid of this menace, but meanwhile, Oedipus' other daughter, Ismene arrived with news from the oracle that a blessing would be instilled upon those who took Oedipus in, dead or alive. King Theseus then gave him refuge, but not after much antagonism from the

Theban group: Creon and Polynices (Oedipus' son).

Then a thundering summoned Oedipus into the sacred grove. Theseus, Oedipus and his two daughters entered. The girls gave him a ritual bath and he said goodbye to them. Then he and Theseus entered the grove. Suddenly, the voice of a god was heard:

"Oedipus, you tarry too long; now is the time!"

And with that, Theseus was seen, alone in the grove, with his hands to his eyes, staring in awe at something unknown to the others. And with this, the life of Oedipus ended.

❁ ❁ ❁

The Oedipus Story and The Logos Bioculture

As you may recall from previous chapters, the logos bioculture is born out of what we have termed a left brained mimetic bioculture, both having their emphasis in the ideographic act of thinking, as opposed to image-making. What distinguishes the two (mimesis and logos) is that in the former, ideography and mythopoesis co-exist as two different albeit legitimate forms of knowledge; in the latter, mythopoesis is completely absent as a legitimate method for the acquisition of knowledge. The myth of Oedipus delineates the parameters of this new knowledge and also sheds light on HISTORICALLY, how this new knowledge gained a stronghold in the ancient world, and also how it has carried over into modern society. In the following paragraphs we shall focus on several elements in the Oedipus story to elucide the birth of this new way of knowing, the event which created the logos bioculture.

If there is anything starkly obvious about the logos bioculture and its heroes, it is the lengths they go through to relegate mythopoetic knowledge to an inferior position, indeed to remove it entirely and replace it with conceptual theorizing. One has to wonder how and more importantly, why this is. The answer is probably that the new knowledge is a form of creation that men and women can both perform, unlike in mythopoetic creation, where the highest form of image making-the creation of another human-was perceived to have been the sole work of woman. Thus, the real reason for the onslaught of the logos bioculture type was to remove sole knowledge from the realm of women and usurp the perceived power of this knowledge for anyone, including men. This act can readily be accomplished through the habitual reinforcement of substitution systems in the human mind, to the exclusion of imaginative, mythopoetic exercises, thereby setting up a "false knowledge" based upon theories external to the human's realm of bodily verification. Once this is accomplished, by constantly shifting the input from the outside, the human remains subservient, ignorant and frankly, dependent on the "knowers" for their entire existence. (This is the philosophy of mercantilism in a nutshell, a fact that shall become quite relevant in a few minutes. And if you doubt the veracity of this explanation, try to get your children to wear clothes other than the brand that they "need" to wear to be in style at school.)

So how did we become so conditioned to reacting solely to things from the "outside'? When we re-created the world through the eyes of our left brain to the exclusion of our other brains, and with that exclusion, of our bodies as well. To understand how and why this happened in our human history and biology, we need to look back at the myth of Oedipus to learn the answers. In the telling of myth in the previous section, certain phrases were highlighted. Let's go back to these now and analyze them in an attempt to rediscover the map of this mindset.

You should recall that Oedipus was born from the lineage of King Agenor of Phoenicia and that his ancestor, Cadmus, founded Thebes and brought with him there, the knowledge of the alphabet. This is a very important point, not because of the alphabet, per se, but because of a forced technology that learning the Phoenician alphabet had on the mind of the non-Phoenician. To give you a modern example by analogy, consider the following: Suppose we have a symbol, ⊙ which stands for "sun" (pictogram) or "son" (phonogram). This connection makes sense in English, in a way, because of the affinity of sound in the homonym; therefore, as a native speaker of English, I could conceivably derive contextual understanding of either homonymic sound (son or sun) when seeing that sign in a piece of writing. But what if I now transplant that same image into Spain and make it mean both "sun" and "son"? Considering that the Spanish word for "sun" is "sol" and for "son" is "hijo," the Spanish speaker would have to learn by rote the connection between the symbol and both words. This act sets up the mechanism in the brain for logos knowledge: knowledge based on conceptual substitution grounded on arbitrary correspondences. There is no way that anyone in this circumstance could use mythopoetic knowledge to apprehend the connections themselves. And this is precisely what happened in many places in the ancient world where the Phoenician alphabet was introduced; especially in Greece. Of course this mindset was already hardwired in the Phoenician cultures by this time, not by this occurrence but by another which we shall discuss in subsequent paragraphs.

The next curious event in the myth of Oedipus is this constant travel to and from the oracular centers in order to get information. An oracular statement is similar to a riddle; the words are uttered but it is up to the individual to figure out the meaning of the words. The utterances are always true; the interpretations by the receiver are not and ancient history is replete with stories based on misreadings of oracular statements. Basically, oracles serve the same function as the voice of demiurge gods do; they are the outside voice speaking to the conscious individual; the right brain talking to the left brain, but since the left brain perceives only the conceptual data, it thinks the voice is from "outside"; not from within itself. And this is the second component of the logos bioculture; the decision on the part of the individual to focus on that oracular voice and choose a mode of action either for or against the message in the voice. This choice is not a choice, for once an individual falls into the "x or not x" mode in regard to that voice, he/she is doomed to fall right into the logos loop. The alternative, when faced with the voice, is to test the veracity of the message by reinterpreting it through the other brains and naturally through the body in an effort to understand the origin of the message. If the message is verified through the bodily sensorium, heed it because the message is true: it comes from your other brains; if not, ignore it for it truly is outside and not of vital importance to your body's well being. In this way, the human decision to follow or reject the voice will be made through the knowledge derived from the exercise of mythopoetic dismemberment; not ideographic conceptualization. How are decisions made in the logos bioculture? The other way; the way that Oedipus made his: by either accepting or rejecting the message of the voice without ever questioning the veracity of the voice per se. To make the example more contemporary, suppose someone offers you a piece of candy and your proscriptive voice says, no. What do you do? If you say yes or no to the voice itself, you are already falling into the logos loop. Now suppose you say no to the candy, reinforcing the voice inside that says no. What do you use to reinforce your own decision? For those humans who have become members of the logos bioculture, what usually enforces that decision is: a number on the bathroom scale or a new study done on the dangers of eating candy or an FDA warning about the brand of candy you were offered or a newspaper article you just read about candy and heart disease. These reasons, although useful in their own right, all reinforce certain basic beliefs about knowledge that have become universal but in fact are not. They are simply the consequences of the embodiment of the logos bioculture. And what are these beliefs?

1. That there is information "out there" that I should know but don't.
2. That somebody has that information.
3. That I need to learn that information or my life is in danger, or my quality of life will be lacking, or I simply will not be considered useful anymore.

And how do I learn this information? By reading, watching TV, attending lectures etc. in an effort to remain "current" on a variety of "issues." Then I can make "informed" decisions about myself and the world and my "opinions" will be correct ones. So, what's wrong with this picture? This is, after all, how we were all taught in school, right? Unfortunately, right. But the problem is, this method of obtaining knowledge is both fraudulent and insufficient. At best it creates very shallow humans; at worst it annihilates them, as the myth of Oedipus articulates. One wonders what would have happened to Oedipus if he hadn't chosen to respond to the message of that voice . . .

The third element in the Oedipus story worth noting is the nature of the original inhabitants of Thebes: the Spartoi. Remember that these creatures were a race of humans, all male, who were born not procreatively, but were sown from the seeds of the dragon of Ares. Sown from the seeds of the dragon of Ares. This is one of the first and most serious consequences of logos creation: the attempt to attribute it to masculine origins. With very little imagination, one must realize that the Spartoi are men created solely by the seeds (read here, sperm) of the serpent of Ares (read here, *membrum virile*). And Cadmus, together with these "men" founded Thebes. In other words, this culture was created without woman. But one reading this story would certainly have a difficult time trying to imagine these men thus created, wouldn't one? It seems inconceivable and ridiculous, right? The answer is yes, and no, because the elements in the story resemble mythopoetic creation accounts so much: the dragon, the slaying of the dragon by the hero, the creation of a new race of humans. BUT, the interpretation of the images in this story are purely ideographic, as the culture is a literate, ideographic one, so the images are not mythopoetic images, but ideographic symbols of substitution, though the purpose of the story is essentially the same as the purpose of other hero stories: to recount the foundation of a new mindset, which in this story is theorizing. So in this story, the serpent is not the cosmos of touch, but the ideographic symbol of the male member; Cadmus is not a hero, but the demiurge god, creating the world of forms in the only way any male god could do it, by masturbation. And the Spartoi are his creations. What's totally unique here is the complete lack of feminine input, except for the cerebral advice given to Cadmus from Athena (the virgin warrior goddess of wisdom) about how to create these "men." In other words, what Cadmus gave birth to here are "head" men: concepts. And with these headmen and his newly acquired wife, Harmonia, he settled Thebes. Of course Harmonia had a history of her own, being an illegitimate child of the goddess of love and the god of war; with the husband of her mother, Hephaestus, the supreme artisan of Olympus. It's no wonder that Cadmus' family was doomed from the outset since the only hope for creativity in a logos society would be through craftsmanship or cleverness, and the Thebans seemed to be lacking allies in at least one of these areas.

While we are on the subject of Cadmus' unique method for populating his city, we must recall that it was the grandson of Cadmus and father of Oedipus, Laius, who seems to have taken literally this method of creation, for he, we are told, was the first man in all of Greece to have become enamored with another member of his own sex, Chrysippus. And also remember, that THIS FACT is what unraveled the house of Thebes, ultimately, because Hera, goddess of marriage and procreation cursed his lineage for this act. So what was it about this act that caused such grief? The act, per se was not the problem. The fact that Laius found this attraction natural, and acted upon it, and refused to lay with his wife, and was willing to SACRIFICE his own son to appease the goddess without consideration for any other way of life was the problem. In short, Laius had eradicated from his mental system the mythopoetic mindset, as evidenced by his blatant disregard for his kingly obligation to his wife, choosing instead to BELIEVE THE VOICE of the oracle, without ever making the connection between the oracular utterances and his own actions, the reason for the oracular utterance. Rather than offering to sacrifice his child in order to save his own fate, had Laius learned how to "know" (read here, in the Biblical sense) through his relationship with his wife, the wrath of Hera surely would have been appeased and the oracle unfulfilled. But Laius had a self fulfilling prophecy hanging over his head because of HOW he KNEW; not because of anything else. And since, unfortunately even his own wife was a child of one of these "sown men," his son, Oedipus, was probably doomed from the start to repeat the "sins of his father."

Oedipus had to be thrown in a chest and drifted out to sea because he was in fact an orphan; his true lineage was of a man without creative knowledge and a woman whose father was created of man but not woman. And in the myth he was marked of that lineage by his phallic name, "swollen foot"; or else, "the one who knows." And being raised by his foster parents, it was possible for Oedipus to really be a hero on returning to Thebes, in the way that other mythopoetic heroes are made: He could have reeducated Thebes in the mythopoetic mode of knowing (imaginatively, creatively, procreatively) thereby saving the human plan of creation for the gods. But he didn't do that. He made the wrong choice at the Cleft Way; not by going to Thebes, but by choosing to respond to the VOICE of the oracle and its message without considering first, the veracity of the message. But why didn't he question the message's veracity? Because his family had lost the capacity for that type of knowledge a generation before and thus the curse of his father passed on to him.

When the Sphinx posed the riddle, perhaps she was testing to see whether anyone in that city still had the capacity of the old knowledge. I think Hera was counting on Oedipus to save Thebes by slaying the Sphinx, not by entering into a dialogue with her; nevertheless, Oedipus once again responded to the voice, apparently correctly this time and was made king. But was the answer correct after all? For wasn't it the fact that he responded to that voice again that got him into more trouble, thereby compounding his doomed fate? I think we'd have to answer, yes. And what was the significance of that riddle, anyway?

The actual riddle of the Sphinx has its origins in ancient Egyptian myth, where the sun god Ra, is born and grows old during the course of each day, walking in his old age on a cane (hence the third leg.) Hathor, the goddess who is sometimes envisioned as a Sphinx and is usually associated with the Greek goddess Hera, was known as the child of Ra, and also as the eye of Ra. In another well known Egyptian myth, Hathor in her sphinx incarnation was sent by Ra to destroy a race of men, which she did. So what is the significance of this information? First, the Theban Oedipus worshipped the sun god, in Greek, Apollo. Secondly, the Egyptian sun god, Ra gave birth to the universe by masturbation, as the following quote from the Egyptian Creation myth confirms:

> "I produced myself from primeval matter . . . I had union with my clenched hand and my heart came out of my hand. I joined myself in an embrace, poured seed into my mouth and sent forth issue . . . my eye weak because . . . they proceeded from me after I became from one god, three from out of myself."[12]

Since there are some who insist that the sphinx episode in the myth of Oedipus is in fact a late addition, much like the pun of his name meaning swollen foot; one could only predict that the reason for the insertion of these elements is to emphasize the male dominated nature of this community. It is also curious that even the name of the city, Thebes, has its Egyptian counterpart, where Ammon-Ra , curiously enough, was the chief deity, even possessing a great oracular shrine there. (More about this later.)

Another fascinating feature about the Oedipus story is his own punishment: blindness. Why blindness? Some have ventured to say that blindness as a punishment really brings "insight," as in the case of the blind seer Tireseas. If this were true, then at the end of the story, Oedipus' sins would have been forgiven and he would have theoretically lived out a better existence. But he lives out a horrible existence: a blind beggar roaming about the countryside. So is his blindness the same kind as that of Tireseas? The answer is no. Oedipus inflicts his own blindness; Tireseas was blinded by one deity and given "insight" by another in its place. The myth of Tireseas' blindness is as follows: One day he saw two serpent copulating and he struck the female one with his staff. Immediately, he turned into a female. Seven years later, he again saw two snakes copulating and struck the male one. He then resumed his masculine appearance. Because of his unique experience, Zeus and Hera, who were fighting over who derives more pleasure from love, men or women, happened to ask Tiresias. He answered of course, women. This so annoyed Hera that she arranged for him to be blinded by Athena, but Zeus put into Athena's head that she should make up for this by granting him the power of prophecy in its stead. She gave him" a long staff to guide him and a long life and . . . he alone among mortals shall retain his intelligence in the afterlife." Oedipus, of course gouged out his own eyes, not for "insight" but once again, for the purpose of REJECTING the gifts of his deity, the sun, to punish himself for his wrongdoing. The tragic part is that to Oedipus, his wrongdoing was fulfilling his destiny: marrying his mother and killing his father. In fact, though, his wrongdoing was that whether or not he did these things, he FAILED to learn the true lesson of his life: that this "new knowledge" brought about by his family was destructive. And he could have saved his family from the same fate that he fell victim to but he didn't: they all died without ever really "knowing."

A final word must be said about the death of Oedipus. Recalling that he somehow miraculously disappeared into the grove of the Furies after responding to the voice of a god who told him that he tarried too long, one is left to wonder what was truly so great about this man? He set out on his life search in response to a voice, he left his homeland in response to a voice, he dies responding to a voice: none of which were his own. This hardly seems like material for a great hero. Yet, since we, with our logos colored glasses, see him as one who defies the authority "out there": church, homeland, family, and even takes his punishment as a man, want to somehow look up to him for having the guts that we wish we could have. But he is not a gutsy hero. In fact he is a weak, sad excuse for a human, for he destroys his home, gods, family, life, all because of one stupid compulsion: to know. And his story is not one of heroism; but rather, of infanticide, patricide, suicide, fratricide, homosexuality, self inflicted mutilation and misogyny, not to mention civil war and even prejudice. This real Oedipus is indeed no hero of humanity at least; of other things one never knows. No, in the end the character of Oedipus is truly a pathetic creature who fought the dragons of words all his life and never, not ever, stopped to ask himself why he had to know.

Exegesis of the Oedipus Myth Cycle

The True Meaning of "Oedipus"

You may recall from the beginning of the chapter that the clue to the true heresy of Oedipus is actually in his own name, Oedipus. Much literature has been written about the origin of this character's name; however, few have managed to link this name in any significant way to the entire myth cycle, and to the larger issue of the birth of conceptual theorizing, the new knowledge brought to "light" by his descendants. This is what we shall attempt to elucidate now, through a reading of the linguistic puns evident in the mythic house of Thebes, as narrated by the Greeks.

Beginning then with Oedipus himself, one notices that his name is derived from either "oidos,"swollen or "oide" he knows or he sees; and "pous" foot.[13] Thus a simple reading of the meaning of his name might be something like, "he who knows about the (riddle of) the feet" referring to the Sphinx's question; or his name may refer to his bound feet from childhood. These are the usual interpretations one is faced with, yet they together betray an equivocal rendering of the understanding of the entire story. This problem is further compounded by two very significant facts regarding these specific references and their relationship to the meaning of Oedipus' name.

The first reading of his name, the one who knows the answer about the riddle of the feet would make sense ONLY if the specific riddle Oedipus answered were consistently about feet; unfortunately this is not the case. According to Athenasus (X.451F) the riddle is, "There are twins, such that each one gives birth to the other and each kills the other." The answer that Oedipus correctly gives in this version is, "Night and Day." This reference obviously has nothing to do with feet! The integrity of the story, though was not altered apparently with this alternate riddle. Why not?

The second reading of his name, that referring to his bound feet, is likewise a superfluous reading of his name, particularly as the portion of the myth wherein his feet are bound was a later addition to the myth, based upon some earlier rendering of which THIS was the pun. What would that earlier rendering be? And further, is there any way to reconcile all of the above information into a coherent understanding of the Oedipus myth? The answer is, yes.

There is one reading of Oedipus' name, not yet considered, which means something entirely different, something that links both of these interpretations together in a colossal pun. That pun is, "he who has male knowledge." Where does this interpretation come from? An understanding of the true origin of the myth as non-Greek; in fact as Phoenician (sometimes confused as Egyptian). It of course is no wonder that the myth would be of Phoenician import, as Cadmus, founder of Thebes was in fact Phoenician. The interpretation of his name is actually as follows:

Eidos=knowledge or seeing;

Pous=foot or the homonymic *membrum virile* (gr. *posthe*)

That would make his name mean something like, "he who knows the phallic way." But what is this masculine knowledge? And how does he get it? If we are to find this clue by interpreting the story, when did he get masculine knowledge? Did he have it when he answered the riddle of the Sphinx? If he did, what was the purpose of the rest of the story? Or, if he got it after he blinded himself, which seems to be the more sensible interpretation from a literary point of view, what kind of knowledge does blindness bring? In fact, what about this blindness, anyway? We've said earlier that Oedipus blinds himself, unlike Tiresias who was blinded by the gods and given the gift of prophecy instead. Though Oedipus blinded himself, we have no evidence that this type of knowledge was bestowed upon him. And further, what does this have to do with blindness?

It has been suggested that Oedipus' blindness had to do with the other meaning of the word, *eidos*, to see. If this were the case though, his name really wouldn't have any significance at all, for if we know anything about Oedipus it is that he "knew," he was clever, he solved riddles; thus there is a real problem here with regard to his lack of sight. That is unless we consider that blindness itself was another popular euphemism for castration.[14] The association between eyes and phalli could best be seen (no pun intended) in the Egyptian "eye of Horus" which also represents the phallus as the "one eyed god." Thus, perhaps hidden somewhere deep within the myth cycle of Oedipus lies a curious strain of an older, no doubt alien myth from which we can learn much about this peculiar strain of humans.

Continuing then in the story of Oedipus, assuming that his "blindness" could be a euphemism for his castration, could there be any significance to the fact that he performs this act on himself? Is this a necessary element for him to perform in order to free himself from his sin of killing his father and marrying his mother? Was this Oedipus' crime? It seems so, at least to him; therefore, understanding the word blindness to mean castration seems well enough justified, while maintaining the integrity of the rest of the story. In fact this may also account for the later interpretation of his name meaning "swollen foot," for obvious reasons.

Taking the above rendering of his name as a given then, we come to understand the story of Oedipus as having something to do with knowledge through castration. And this seems to be the only coherent interpretation of these elements; otherwise, it is a very bad story with simply too many literary ends left untied, such as what is the purpose of his gouging out his eyes with his mother's brooch? What is the purpose of the riddle of the Sphinx? Why was there need to explain about his bound feet, etc.

If this curious interpretation of the name of Oedipus were an isolated incident, one could always leave it in the realm of possibility; however, when we turn our attention to the other elements of the myth cycle, we begin to see the larger picture out of which our Oedipus emerges as a mere example. Retracing then our steps through the labyrinth of his myth pattern, the history of this "masculine knowledge" becomes blatant. Let us begin with a rudimentary analysis of some of the other names in the myth cycle to enlarge the scope of our exploration.

Poseidon, Father of the Phoenicians

Returning then to the beginning of the myth cycle, we recall that the divine ancestor of the house of Cadmus the Phoenician is the god Poseidon, who fathered the entire race of Oedipus through the two daughters of Epaphus, Libya and Lysianassa. The name Poseidon (gr. *posthe eidon* means masculine knowledge. Plato says his name means chained foot, which we shall soon see is actually a synonymous meaning to masculine knowledge.) is derived from the exact same origins as Oedipus, and like Oedipus, much has been written on the true meaning of his name, with little success.[15] It is no doubt that Oedipus is simply a Greek word pun on Poseidon, perhaps reversed because of the fact that Greek was read from left to right, while Phoenician was read from right to left. The unique inclusion of Poseidon in this myth cycle also serves as a clue to the true history of this myth and indeed leads us right back to the founders of the heresy of which we speak; the ancient Phoenicians.

In order to begin unravelling the rest of the mystery surrounding this secret heresy, let us first say a few words about the god from whom this genealogy comes. According to Greek myth, Poseidon was the child of

Cronus and Rhea, brother of Zeus and the other Olympian deities Hestia, Demeter, Hades and Hera. The myth of the birth of these children relates that their father, fearful that a son would overthrow him, swallowed up each of his children when they were born. According to one tradition, Rhea, when Poseidon was born, whisked the child away to be hidden among a flock of lambs, replacing the boy with a lamb instead, which Cronus purportedly ate thinking it was his son. Thus, the first mythic reference to a substitute victim.

When he grew up, Poseidon shared dominion over the earth with his two brothers, Zeus and Hades. Poseidon was given the sea, thus his association with water. He was also the god of horses and bulls as well as earthquakes. These epithets recall his main functions. He carries a trident (three pronged spear) which he plows into the earth to create water; he "procreates" by mating with sea nymphs to produce children, his wife, Amphitrite, being the best example. Many of his offspring were either giants or monsters (Antaeus, Otis, Ephialtes, the cyclops Polyphemus, the Moliones) or else very cruel humans, most of whom dealt with foreigners by killing them (Busiris, Cercyon, Sceiron, Eryx). (It is interesting to note that many of the people killed by the hero Heracles were in fact sons or descendants of this god.) According to a Libyian myth, Poseidon was the father of Athena, by the river goddess Tritonis. (The alternate version of the myth, which is actually for our purposes the same, is that Athena sprang fully grown and dressed in armor, out of the "head" of Zeus, after the "limping god" Hephaestus slashed Zeus' "head" with an axe. Another castration theme? Remember the "sown men" of Cadmus in that myth, the warriors who sprang up fully grown and dressed in armor, having been created from the "teeth" of the serpent? More on this later . . .) He is also credited with fathering Parnassus, eponym of the mountain on which the Delphic oracle resides, Phocus the Corinthian who founded the land named after him, as well as many of the Argonauts. And finally, we arrive at the lineage of his children by Libya and Lysianassa; a group known to the Greeks as the Phoenicians.

The worship of the god Poseidon, curiously enough exactly corresponds to the migration pattern of the Phoenicians by both mythic as well as historic accounts. Since the Phoenicians were primarily sea farers, it is no wonder that Poseidon would be their patron; thus, he was the favored god of such places as the islands of Crete, Chios, Cos, Delos, and Corcyra and of seaports such as Corinth. Less understandable though from his water associations, is his association with such inland places as Delphi and Northern Boeotia (where Thebes is located) and Attica, particularly Athens, where both Poseidon and Athena share houses on the Acropolis (the Erechtheum and Parthenon, respectively). We shall soon discover that his worship in those places has less to do with his seafaring epithets and more to do with his Phoenician origin, for the chief god of Phoenicia was also worshipped on mountaintops, which explains his giant offspring as well as his mountain cult centers.

One more piece of information about Poseidon that we shall soon discover betrays his Phoenician origin is that he also was known to have love affairs with men. The youth Pelops is one such example worth elaborating upon. King Tantalus, father of Pelops, invited the gods to a banquet. To test their divinity, he killed and dismembered his son Pelops and served him as the main course (for according to Greek myth, their gods were repulsed by human flesh and therefore wouldn't eat it; hence the test). None of the gods ate the meal, except for Demeter who took one bite of the shoulder before realizing it was human flesh. The gods restored Pelops to life, replacing his missing shoulder with an ivory one. But Poseidon, on seeing the beauty of the child, whisked him up to Olympus as his lover. He was later returned and became king of Pisa. It was this king's son, Chrysippus whom the father of Oedipus, Laius fell in love with and, like Poseidon did with Pelops, abducted. Thus the myths about Poseidon always contain some connections, though occasionally remote, to our story of Oedipus. We also find connections between this house of Pelops and Poseidon when we learn of such things as the name of Pelop's wife (Hippodameia,"horse-maiden"), which no doubt refers to Poseidonian influence as well as the name of the particularly Phoenician region of Greece, the Peloponnesus, which is in fact named after him. An Arcadian myth links Poseidon, Demeter and perhaps even the wife of Pelops as well. In that myth, Demeter, while searching for her daughter Persephone was raped by Poseidon. She tried to change herself into a mare to avoid his advances, but he changed himself into a horse and raped her anyway. She then gave birth to horse, Arion. (Pegasus was also sprung from a rape by Poseidon, of Medusa, whom Athena changed into a monster.)

None of the above accounts however show more of a link between the god Poseidon, his Phoenician origin and the myth cycle of Oedipus than those of his hero sons: Bellerophon and Theseus, and of the hero Perseus who, although claiming to be the son of Zeus, was in fact, the mortal son of

Acrisius, ancestor of Belus, the son of Poseidon.

Bellerophon

Bellerophon was the son of king Glaucus ("gleaming blue- green") and Eurynome, the king and queen of Corinth, though his true father was the god Poseidon. Though he was originally named Hipponous ("of horses"), he was forced to change his name to Bellerophon ("slayer of Bellerus") when he accidentally killed a man by that name. As a result of that act, he was banished from his homeland and sent to Tiryns to the court of King Proetus (see above) who purified him. Proetus' wife, Stheneboea ("mighty cow") fell in love with Bellerophon and when he rejected her advances, she told her husband, "You will die, Proetus, unless you kill Bellerophon, who seeks to have his way with me, though I do not wish it." Proetus, fearful of killing the boy himself, sent Bellerophon to his father-in-law's kingdom with a note telling his father-in-law to kill the bearer of the note. King Iobates received Bellerophon and after nine days he was given the note. Since he didn't wish to have the blood death on his own hands, the king sent Bellerophon on a mission to slay the mighty monster that was ravaging his country, the chimaera (a fire breathing beast with a lion's front, goat's middle and serpent's tail), certain that the man would be thus killed.

Bellerophon went to the seer Polyeidus ("knower/seer of many things") to ask for help in slaying the chimaera. He advised the man to sleep for one night on the alter of the goddess Athena. That night he had a dream where Athena gave him a golden bridle with which he could saddle the famous Pegasus, the winged horse of Olympus, and thus successfully slay the beast.

She also told him to sacrifice a bull to Poseidon so that that god would aid him in his fight. When Bellerophon woke up, he saw the most miraculous sight: the golden bridle of his dream AND the horse Pegasus gently drinking from a nearby stream. The young man immediately harnessed the horse and went off to find the chimaera, which he did and slayed.

When King Iobates learned of the success of Bellerophon, he attempted to have him die three more times, by fighting various humans; all to no avail. In the end, the king realized that this man must truly have been a god's son so he divided his kingdom with Bellerophon and gave him his other daughter as a wife, who bore him two sons and two daughters.

Not content to live a mortal life, Bellerophon one day decided that he should live on Olympus with the other gods, so he leaped onto Pegasus and began his ascent to the heavens. Zeus, on seeing Bellerophon's hubris, sent a gadfly to sting Pegasus, causing Bellerophon to plunge back to the earth. As a result of the fall, he became both lame and blind. He lived out the rest of his life as a wandering begger, always cursing the gods. His children were all killed by the gods and no one ever recalled the place or manner of his death.

Theseus

King Aegeus ("goatskin") of Athens was childless, though he longed for an heir to his throne. When he went to Delphi seeking oracular advice on the matter, the enigmatic message was, "loosen not the projecting foot of your wineskin until you reach Athens." This message was undecipherable to the king, so on his way home he stopped by Troezan, the region of King Pittheus (son of king Pelops, above) to ask for an interpretation. Though he didn't tell

Aegeus, Pittheus knew exactly what the meaning was; in fact, it had a double meaning: do not get drunk AND do not mate until you return to Athens. Pittheus decided to cause the king to disobey, albeit unknowingly, the oracular warning, so he arranged that Aegeus got drunk and that he slept with Pittheus' daughter, Aethra. What the king didn't know was that his daughter had been raped by Poseidon that same night, while she was performing a sacrifice on a local island. When Aegeus left for home, knowing that Aethra was with child, he told her that if the child were male, send him to Athens. The king left with Aethra two tokens, his sword and his sandals, which he placed under a huge rock, that if the child could retrieve them, he would surely be worthy of his father's acknowledgement.

Needless to say, when the young boy, named Theseus grew of age, he was able to lift the rock and retrieve the tokens, so his mother sent him on his way to Athens. On his way there, Theseus overcame a series of monsters and beasts:

- Periphetes (a crippled son of Hephaestus who cracked people's skulls with his bronze club). Theseus killed him and carried his club as a trophy.
- Sinis (the pine bender, who tied his victims to two trees, head on one, feet on the other, and then let the trees go, thereby splitting the poor victims apart.) Theseus performed this same act on Sinis.
- Phaea (the wild sow of Crommyon who was ravaging the countryside.) Theseus killed it and went on his way.
- Sceiron (a robber who lived on the rocks along the isthmus, who asked

to wash people's feet, and then threw them off the rocks to be eaten by a giant sea turtle.) Theseus performed the same act to Sceitron. King Cercyon (who forced foreigners to wrestle him, killing them either during or after the match). Theseus wrestled him to his death.

- Damastes/Polypemon (who invited weary travelers to sleep in his bed and then either cut them or stretched them to make their size fit his bed exactly. He was also known as Procrustes, "the stretcher.") Theseus stretched him to his death. When he finished his exploits he was purified for his killings at the temple of Zeus Meilichus, and arrived at Athens a hero, having killed all of the isthsmean bandits.

In the seventeen years between the birth and return of Theseus, King Aegus of Athens had had another son, Medus by the sorceress Medea, so when the young Theseus arrived, though Aegeus didn't know it, Medea recognized him as the king's long lost son and heir who would outplace her own son, so she tried to have the young man poisoned at the banquet made in honor of his arrival. Luckily at the last minute Aegeus recognized the tokens and accidentally struck the poison cup out of his son's hands when he was reaching out to embrace him.

Theseus lived in Athens as a prince, continuously helping the King out in his various difficulties, the most impressive of which were the slaying of the wild bull at Marathon and the slaying of the minotaur, the half man half bull monster of Crete. It was after this expedition that his father, Aegeus died a tragic, mistaken death. You see, the father and son had agreed upon a sign to let the king know whether Theseus' exploit had been successful or not: a white sail meant he was victorious, a black sail that he had died. By accident the wrong sail went up on the victorious home voyage, thus King Aegeus leaped into the sea and died. The Aegean Sea is named after him. On his return, Theseus met with the twelve scattered communities around Athens and convinced them to join together in one single city, which they did, and he ruled them as a benevolent father. It was also at this time that he granted Oedipus his final resting place.

He is said to have married Ariadne ("the very holy one") of Crete ("sister" of the minotaur) but abandoned her on an island shortly thereafter . (She eventually married Dionysos.) He raped the Amazon queen Hippolyta and bore a son by her, Hippolytus, and finally married the Cretan princess, Phaedra ("the radiant one").

The relationship between Phaedra and Hippolytus was less than positive, as Phaedra, upon seeing the youth fell passionately in love with him. When the boy neglected her advances, she told Theseus that he tried to rape her, whereupon Theseus called upon his father Poseidon to kill his son. A bull came out of the waters and Hippolytus, who was riding his horses was dragged from his chariot and killed. Phaedra committed suicide.

On his return to Athens from Troezan, where he had been placed in exile for one year because he killed a man, Theseus was met with opposition by some who accused him of being a dictator. He was defeated and therefore cursed the land of Athens and went into exile and his children were likewise expelled from the city. He went to the island of Scyros, where his father owned some property, but one day, while walking along a hill, he fell off a cliff and was killed. Some say the king of that island pushed him; nevertheless, this is how he met his end. Some years later, after the Persian Wars, the bones of Theseus were returned to Athens, as the oracle of Delphi had instructed the Athenians that they should revere him there. This oracular statement was obeyed because some soldiers reported that it was indeed the ghost of Theseus who had helped them in the battle of Marathon.

Perseus

Perseus ("to cut") was the son of Zeus and Danae, though many claim his true father was his mother's uncle, Proetus (see above). As you may recall Danae's mother was locked up in an underground chamber because her father Acrisius received an oracle saying that if she were to have a son, that child would kill him. That child was indeed born, and was named Perseus.

King Acrisius, on discovering the child had been born, took him and his mother Danae and locked them in a chest and flung it into the sea, so that they both would be drowned and he would be rid of the threat of overthrow forever. The chest landed on the island of Seriphus (Serapis?) where it was found by a fisherman, Dictys. Dictys ("netman"), upon hearing the story of Danae and the child took Peresus to the temple of Athena to be raised. Danae meanwhile was attempting to ward off the numerous advances being made to her by Polydectes ("he who receives many"), brother of Dictys and also the king, for she had no intention of having him defile her. Polydectes, realizing that should he rape Danae, her son would seek revenge, sought for a way to rid the world of Perseus so that he could possess Danae for once and for all. He arranged a feast for all the noblemen of the island, knowing that each would have to bring him a customary gift. The day before the party, the king asked Perseus what he intended to bring. Perseus answered, "whatever you'd like me to bring." The king asked for the head of the gorgon Medusa. The next day, all the other noblemen and even Perseus came to the party bearing gifts of horses. The king however refused Perseus' gift, demanding the head of Medusa and telling the boy that he would rape Danae if Perseus went back on his pledge. Thus, the young Perseus went out in search of the Gorgon's head.

Hermes and Athena aided him in his quest. Hermes first led him to the Graea ("grey sisters"), three old women who shared one eye and one tooth among them, so that they could tell him where he could find the gorgons. They also told him where to find the other elements he needed for his venture: a cap of invisibility, winged sandals and a leather knapsack, in which to keep the gorgon's head. Hermes also gave him a sickle shaped sword with which he would cut the gorgon's "head." Armed and ready to go, Perseus set off.

Medusa was the only mortal among the gorgons, she herself having been turned into a monster with snakes for hair, by Athena when that goddess discovered she had been raped by Poseidon. Out of Medusa's neck from the rape were sprung Pegasus the horse and Pegasus' twin, Chrysanor, "golden sword." Now Perseus was well aware of the fate of anyone who looked upon this creature: they'd be instantly turned to stone. So Athena concocted a way whereby Perseus could guide his hand and lop off the gorgon's head without actually looking at her. He did it by using the reflection on his polished shield. When he finally cut off the head, he threw it into the sack and began his journey home.

On his way home, Perseus came upon a frightful sight; a beautiful young woman chained to the side of the sea about to be eaten by a huge seamonster. The girl, Andromeda was being sacrificed to Poseidon because her father had heard from the oracle of the god Zeus Ammon that, because his wife boasted that her daughter was more beautiful than the sea nymphs, it was Poseidon who sent the monster to ravage his kingdom. Perseus, in attempting to save the girl, was completely eaten up by the sea monster, but he managed to cut his way out of it and save both himself and the girl. The king then gave Andromeda to him in marriage and they set off for Seriphus.

On arrival they encountered a horrible sight; Dictys and Danae were clinging to the sanctuary of Athena while Polydectes and his men had surrounded the place in order to capture them. At that moment, Perseus revealed the gorgon's head, and all the men were turned to stone. Dictys then assumed the throne and Danae migrated to Italy and there settled a city.

Perseus and Andromeda then set sail for Argos to claim his throne. King Acrisius, hearing of Perseus' intentions, fled to the island of Larissa in fright, but Perseus followed him there, insisting that he had no intention to retaliate or harm the king. Thus they reconciled and that night the king invited Perseus to participate in the local athletic games being held. Perseus agreed, choosing to hurl the discus. Unfortunately, on his first throw, he accidentally killed the old king, the discus landing either on his foot or his head, and Perseus thus inherited the throne. Since he was embarrassed to rule in the city whose king he had killed, he switched kingships with his cousin, and ruled Tiryns instead.

Perseus met his end as king though when he opposed the cult of the god Dionysos, which had spread into Tiryns from neighboring Thebes. Though Perseus succeeded in killing many of that god's followers, he was ultimately defeated and had to flee back to Ethiopia, home of Andromeda. There they had many children, especially a son, Perses who founded the Persian empire by gathering up all the scattered tribes of the area together under his rule. Perseus and Andromeda, upon their deaths became constellations.

Of Heroes and the Moon

As one can see from the above examples of hero myths, stark similarities and parallel motifs begin to emerge both among these myths themselves and as they specifically relate to the Oedipus myth. One can therefore safely assume that the myth of Oedipus is not a unique depiction, but rather a composite of or else a re-telling of an earlier myth cycle. In attempting to unravel the origin of this myth cycle and its meaning, though, one becomes seeped in a hopeless morass of names of characters and places. The repetition of certain key elements becomes evident though, when examining the meanings of the names of certain characters as well as the almost formulaic repetition of actions that these characters perform in each cycle, such that, through a different reading of the myths, one can safely determine that all of these myths are indeed merely a repetition of one profound original myth, perhaps shared by many ancient peoples.

The rudimentary characters, in addition to the heroes themselves, in all of the above myths can be summarized as follows:

-Bulls (minotaur, bull of Marathon, Zeus' mating with Europa in that form, Bellerophon's sacrifice to Poseidon, Theseus' curse on his son . . .)

-Monsters (chimaera, isthmian bandits, sphinx, minotaur, sacred serpent of Ares, sea monster, Medusa . . .)

-Seers/Oracles (Delphi, Zeus Ammon, Tireseis, Polyeidos . . .)

-Old Kings (Proetus, Aegeus, Laius, Acrisius, Polydectes . . .)

-Cows (Io, Hera, cow Cadmus followed to found Thebes . . .)

-Queens (Io, Jocasta, Phaedra, Sthenoboea, Europa . . .)

Actually, there are really no other characters in any of these stories and the basic plot seems to be something like this:

An old king is warned that his wife/daughter will have a son who will overpower him so he tries to stop the prophecy from happening, but it happens anyway and a son is born who kills him and takes over his throne, but not for long . . . This son, before he becomes king, does something with one or many monsters. He gets married to a close relative of the old king (usually) and has kids but the kids don't succeed him. He eventually has to leave his new kingdom and he dies some odd way.

Of all the characters in these myths, the ones whose true nature are most obvious, when looking at the meanings of their names are the queens; therefore, we shall begin unravelling this seemingly enigmatic cycle of myths using them.

In short, all the queens in every one of these myth cycles maintain lunar or bovine appellations: Io "moon"; Europa "full moon"; Jocasta "shining moon"; Sthenoboea "mighty cow"; Phaedra "radiant one." Even the goddess

Hera's most famous epithet means "cow eyed" and the land founded by Cadmus was called "Boeotia," cow country. This, coupled with the obvious other bovine references above, such as this family's ancestor, Io's transformation into a cow, reveals that the lunar cycle seems to play a significant role in deciphering this story. Remember also that regardless of what actions these queens perform, their names remain lunar/bovine. The bovine plot thickens when we realize that the founding of two cities in our myth cycle, Thebes and Carthage, were founded by sacrificing cows-especially because the Phoenician word for city," byrsa," sounded to the Greeks like their word for cow's hide. (This may explain the stories about how the Phoenicians measured the parameters of their cities by the length of a cow's hide!)

Turning our attention next to the male characters in the myth, we begin to realize that unfortunately, no similar affinity can be discovered, though there exists some evidence to conclude that perhaps at one point there would have been some reference to the solar cycle in these names; hence, for example Cadmus means, "the one from the east," perhaps an epithet for the sun. Some characters seem to be affiliated with horses (Hippolytus, Hipponous); others, to this "masculine knowledge" (Poseidon, Oedipus, Epaphus, Chrysippus) and still others are merely referred to as kings or rulers (Polydectes, Laius). Finally, there is this class of male characters with a uniquely semitic ending to their names (Theseus, Perseus, Aegeus, Zeus), which other than that seem to share little in common with each other.

The monsters likewise seem to appear in the myths merely as obstacle formulae in keeping the hero from his rightful position, though other than that, they share nothing in common.

Finally, the role of the seers and oracles all seem to show merely one thing; though the information given in the messages are always true, the interpretation is entirely dependent on the readers, most of whom misread the message or try to avert it.

But what about the bulls? How do we reconcile the familiar appearance of the bull in each myth cycle, albeit in different places? Perhaps understanding this element is the key to understanding the entire story.

One truly baffling fact in analyzing these myths as a coherent whole is that the bull image doesn't remain constant. At times, the bull is the male god, Zeus who mates with mortal women in the form of a bull, as in the cases of Io and Europa; or simply with the god in his bull form mating with his divine consort Hera in bovine form. At other times, the bull is a sacrificial animal which the hero slays, as in the case of Bellerophon; or which the hero slays as a monster, as in the case of the minotaur. Occasionally the bull is a benevolent, divine assistant to the hero, as in the case where the bull of Poseidon leaps out of the water to kill Hippolytus. In fact, the bull is substituted in almost every facet of the male characters in these myths, though not in any seemingly consistent way. The obvious question is why? The answer has to do with the mind/mind split and this is the crux of our analysis. That bulls were popular subjects of ancient myth and art cross-culturally is generally evident; what meaning is ascribed to the various cultures' renderings of bulls is divergent and herein lies the beginning of our analysis of the mind/mind split and the heresy of Oedipus.

The image of the bull as the male consort of the cow has its parallels in most ancient myth systems: in Phoenicia, the god El mates in bull form with the goddess Asherat; in Ugarit, Baal mates with the heifer goddess Anat (or Aliyan, son of Baal mates with his sister Anat as a heifer); in Egyptian Memphis, Apis the bull god was conceived when the moonbeams fell on a cow in heat; later, Osiris was personified as a bull with his consort Isis, as cow. But perhaps the most intriguing account of the bull and cow motif has its origins in the creation hymns of the Rg Veda[16], the earliest texts of the Indo-Europeans, and as we shall soon discover, this rendering is perhaps the original of which all the preceding myths are the copies. By elaborating upon this creation myth, it will become evident how:

1. The myth evident in the Rg Veda is the original out of which the others were created.
2. Within even the Rg Veda, there was a struggle for supremacy between two interpretations of this myth .
3. The ascension of the bull/cow motif was brought forward by the priestly class in the WRITTEN texts of the Rg Veda (they were originally chanted.)
4. As a result of the written supremacy of the bull/cow motif, with its primary supporters being the warrior class and later the mercantile class, a shift in emphasis occurred placing human procreation at a secondary position to masculine creative arts, INCLUDING the art of writing (hieros logos.)
5. Through trade routes and military expansion and the availability of

> literacy through the creation of the alphabet (which eased the heretofore burdensome task of writing), this mindset gained a stronghold in the ancient world, thereby relegating to a virtually non-existent position, the roles of first women, and then ultimately even the gods.

What is of paramount importance for us, though is that in this original Indo-European creation myth under discussion, the bull is not exactly consort of the cow, for preceding the actual existence of the bull or the cow was the bull-cow "Visvarupa," a pre-creation hermaphroditic essence of both creatures contained in a common sphere imagined as interconnecting paired bowls (which ultimately became identified with the vault of the sky and the earth, hence the terms sky god/earth mother.) But preceding this form, they were part of the original One, the "creator of form-causer of form to be-all form that could exist" One, ("Tvastr-Satvitr-Visvarupa"). This One, formless as yet, existed in the watery darkness of a maternal matrix. Thus, a complete understanding of the myth, as we can see, requires the creative powers of both primordial sexes BEFORE there was such an understanding of division; however, in later renderings EVEN of this original myth, this understanding was obscured by a misreading (an ideographic misreading) of the nature of the bull who sprang from the bull/cow. Let us begin our analysis first by describing the complete myth and then later, by portioning off the (mis)interpretation of the bull usurper to place our Western myths of the bull and cow in their true light.

✿ ✿ ✿

Creation Before the Beginning (Rg Veda)

The beginning of creation occurs with movement, (remember maia?) with the One breathing, thereby causing an ebb and flow of the watery matrix. This oscillation culminated in the heat of desire, and with desire, being arose in the One, causing expansion. Thus the One extended downward as the pillar of the universe. The name of this pillar was "one foot unmanifest." This pillar ("Aja-Ekapad") though had no part in forming; for "forming" and "driving to movement" were still in the "bullcow"; the One about to be made manifest. Now this locus, moment, event, instant, is the source of the inner seeing of the "seers"; a "sight" that comes before manifestation , before opposites, before the world of forms. ONLY after they are self revealed, are they then seen as the cosmic opposites of bull/cow; male/female; night/day; sun/moon. Then the bull takes the image of the sky father and the cow, as earth mother. And the line that seemingly separates them is the pillar of the One when they are apart in existence; though when their halves join together, they return to the unity of the origin.

The pair that assure the continuity of the cycle of coming together and going apart are the heavenly twins, the Asvins. They are the Rg Vedic heroes whose name means "having horses," for their dual yet singular function is to guard the crossroads of those moments before unity and after that which has just become manifest; they are the equilibrium of the cosmos, assuring that the eternal flow remains perpetual. They are likewise the two simultaneous ways of knowing: all together and separated.

With these few Indo-European references as a start, we can begin to make certain parallel interpretations for at least some of the names of the male characters in the Oedipal myth cycle; we know that the bull refers to both the sky father as well as the masculine procreative forces in general and that the horse references probably originated in the original hymns of the Asvins. These analogies so far seem to hold true, at least as far as the nominal associations are concerned. Now let us consider the specific roles of other twins of the Indo-Europeans, an idea curiously expressed also in the myth cycle of Oedipus (Agenor/Belus, Danaus/Aegyptus, etc.), for we shall discover that, although the idea of twin brothers appear in the Oedipus myth cycle, their original meaning seems to have totally lost significance, for reasons which will soon appear fairly obvious.

The Rg Vedic twins, exemplified by such pairs as the Asvins, Yami/Yama and before them, the twins "rushing waters" and "bullcow" recall TWO ways of knowing; one way is how the seers see (all things together before the separation of creation); the other how the other ones see (as separate entities of things: bull, cow, male, female, up, down, etc.) Thus one twin was always a child of the gods, while the other was always a mortal, because of the two ways of knowing. Mortal here means that it is subject to movement from coming into being to going out of being, while immortal is the realm only of the One before separation.

In the myth of the twins Yama and Yami, children of the rushing water and the brightness which causes brightness, who are also born before creation, the female twin whom we shall call Yamayami, since she also carries with her her male twin, yearns to mate , but the male Yama part alone chooses rather to remain separated, coming only together by choice. Now since he created choice, he broke the unity of their beginning (as Yamayami), thus causing himself (the Yama alone part) to have to die; his body being the mortal evidence of his creation of choice. Since however he is also the divine force of love, as the masculine force of generation, (when he is Yamayami), he is also immortal. So when he is in a body, he is mortal; after he dies, he becomes immortal, the King of the path of the fathers (as opposed to the path of the gods), a pathway that he himself created by his choice. The other path, the path of the gods, is left for women and some seers and a few other mortals eventually.

Yamayami were born from the union of the female "Saranyu," meaning "the rushing of waters" (emphasize the rushing not the waters) and the male "Vivasvat," meaning "radiance." Saranyu was the sister of the bull/cow, as both were children of Tvastr, the part of the One that is the form maker. (But remember, as of yet no form has been made. All that you have read so far still exists in the realm of the non-existent, in potentia, if you will.) From Saranyu and Vivasvat, the Asvin twins were created and the twins Yamayami. But because of Yama's choice, the god's removed Saranyu, replacing her with an image of her, since Yama made such a thing possible when he separated himself from Yamayami, thereby creating a space in creation for an image matrix. They called the image of Saranyu Savarna (like One) and she mated with Vivasvat creating Manu, the ancestor of the human prototype. So the mother of man is a manifested image of feminine power and the father is divine brightness. All of this occurred because of Yama's choice to become apart at times from Yamayami. But Yama's reason for coming apart was so that the gods could be known, for if there were no knower, how would they ever be known? So because of Yama's choice, the mortal world was created. Then, you may ask, what about the realm of the bull/cow? Unfortunately, just like Yama, the bull/cow, who as you remember was the part of the One representing "all form or mere form" fell down by HIS own choice (which was a different choice from Yama's) and was destined to leave the eventual realm of the gods, even before creation manifested itself. The following section explains how and why that occurred.

The Myth of Visvarupa (bull/cow)

To recap a bit, Visvarupa, the bull/cow, was part of the original unity inside the One which , besides the bull/cow included Tvastr (the former) and Savitr (the impeller). Since bull/cow was only formed (again, not ACTUALLY formed, more like the matrix for form), his father was Tvastr, the "former." Now while Saranyu flowed forth and gave "birth" to the Asvins, Yamayami and (sort of) Manu, Visvarupa never had progeny because HE refused to give up the powers that were contained in himself. So as a result, "mere form,"the son of "all form Visvarupa" (which is "the forms of things, Visvarupa" plus the other two powers of "impelling, Savitr" and "forming, Tvastr") had no place in the realm of existence prior to creation. In order then for mere form to become manifest, the god Indra[17] had to slay Visvarupa, (in fact, has to continuously slay him) but nevertheless, mere form still had lost its place in the uncreated realm, for "creation of forms to be" as distinct objects were released only by the force of a creator god outside of the One, the god Indra, thus the objects as created are really not part of the source from where creation comes, but are only "created" by Indra, the one who cut open the bull/cow and released the powers of creation that were held in the cow part of bull/cow (before they were separate of course). And by doing this act, Indra took over the image of the bull since he was male AND the powers of the cow since he is the one who released the forms, not the cow; thus Indra became god of MERE FORM ONLY. Now since the world as perceived by humans living in the world is only of mere form, Indra likewise appears as the creator god TO THOSE HUMANS WHO FOCUS ONLY ON FORMS. But since the world of mere form has nothing at all to do with creation in the cosmic sense, to the entire plan of creation, Indra is merely an appendage that accounts for the rejected realm of mere forms (read forms unable to create) brought about by the refusal of Visvarupa to procreate. But luckily, humans and the gods perform intercourse in all the other realms, thereby allowing the divine pattern of creation to reveal itself.

❁ ❁ ❁

Bulls as Masculine Prototypes

We have seen from the above myth of creation that there appear to be several references to bulls, not all necessarily representing the same act; a similar situation to the way bulls were portrayed in the Oedipus cycle, namely, in varying ways. What we can begin to notice though by looking further at the Indo-European myth is that the images of bulls are consistently dissimilar in the same way that the bulls in the Oedipus myth cycle are. In other words, the acts of bulls are varied, but these varied acts are consistent among the myth systems. The only time that a problem arises with interpretation of these prototypes is when one views them by ideographic "seeing" and not by mythopoetic "seeing." Recalling for a moment the previous myth cycle, one realizes that these two ways of seeing were actually inherent in that cycle; the way the seers see (mythopoetically, as One) and the way other mortals see (ideographically, as disparate entities). In the former case, one would receive a mythopoetic interpretation of the narrative; in the latter, an ideographic one. In this way then, each of the prototypes always yielded two interpretations; one mythopoetic and the other ideographic. The problem arises, though when only the ideographic interpretation is exaggerated and the mythopoetic interpretation is kept mute. This is what happened with the original Indo-European myth when the story was transported via the Phoenicians; the ideographic story became THE story; the mythopoetic version fell by the wayside into the realm of the unknown, until now.

In this section, we shall first delineate the masculine prototypes of these myth cycles, then later, distinguish their interpretations, both mythopoetically and ideographically. And finally, we will show, by the linguistic criteria available, how the phonemic shift among cultures translated into the cognitive shift among humans, the mind/mind split, and then finally, how the Oedipus myth finally crystallized this interpretation in the West through its acceptance in ancient Greece.

To begin then, there are basically four masculine prototypes inherent in the image of the bull, (and one of the ox, which shall be discussed in depth in another section), and depending upon what the bull DOES, one can place the corresponding image into the masculine paradigm. In both the Indo-European myth cycle, as viewed in the hymns of the Rg Veda (above) and also in the Oedipus myth cycle, the roles of the bull can be divided as follows:

1. Bull as half of the Bull/Cow duality
   (Zeus mating with Io, Hera, Europa; Rg Vedic "Visvarupa")
2. Bull as Masculine Procreative Fluid
   (Poseidon "creating" by striking his trident to create waters; Poseidon mating with sea deities; Rg Vedic Indra who "creates" by hurling his weapon at "Visvarupa")
3. Bull as Symbol of Ruling Leader's Continued Virility
   (When Theseus calls upon the bull of Poseidon to kill his own son Hippolytus; in Rg Veda when Yama takes on the title Samana: "the leveller," god of death , the bull headed, blue monster)
4. Bull as Hero Usurper
   (When Bellerophon sacrifices a bull to Poseidon; in the Rg Veda when a bull sacrifice is made to Indra for success in the male's next endeavor-to usurp someone else's power.)
5. Ox (to be discussed later).

Bull as half of the Bull/Cow duality.

In the Indo-European myth, the two ways of knowing the bull in this capacity are as "Visvarupa," part of the One which is entwined as bull/cow and alternately, when the elements of duality are separated, as bull and cow, light and darkness, sun and moon, etc. Thus, the mythopoetic interpretation of the bull/cow union is an expression of the return to chaos, the beginning before form, in an effort to return to the source and create again. In procreative terms it is the *hieros gamos*, the holy union that causes life to burst forth. So, for example in the myths of Zeus and Io, or Zeus and Europa or Zeus and Hera, progeny are born to the couple thereby creating new ancestral lines of humans. Conversely, the ideographic interpretation of the bull here represents ONLY one half of the duality of creation; the half which is bull, sun, male, bright, heavenly, father, etc. The female part then, becomes cow, moon, dark, earthly, mother, etc. (One doesn't have to interpret much to see which gender receives the short stick with this analysis!)

Bull as Male Procreative Energy

In mythopoetic terms, the male "uncreated one foot" of the Rg Veda exemplifies the masculine procreative energy as the pillar of the universe that breaks apart the One from the maternal waters. It is the force that brings about differentiation, BUT it has no part in actual creation itself. And when the One is in the position of union, it does not exist as a force of separation; it only appears when separation is about to manifest. It is also, remember a PART of the One. In ideographic terms, Poseidon and Indra respectively personify the Greek and Indo-European deities of male procreative energy, as the forces that break open NO form, and make form. They therefore become the creator deities of the material world, without any mention of any feminine counterpart, for in this interpretation the object which they break open (here no-thing) would have been feminine, had they given a name to what they were breaking open, but since they only create forms that have no divine essence, no feminine counterpart is needed. BUT it appears that this male procreative power has an actual capacity to create alone; in fact, it does not, it only creates the semblance of things, "mere forms" as the myths say. Nevertheless, this male creator becomes the model for the divine artisan who appears as omnipotent deity in later history. He is also the model for homosexual union as a source to a higher form of creation than procreation, namely the creation of ideas, as emulated by certain Greek philosophers and later (male) intellectuals of the word.

Bull as Symbol of Old King's Continued Virility

In mythopoetic culture, movement is primary, thus any culture that remains stagnant either by dogma or impotence is bound to fail. In myths where the king is able to "call up" his masculine powers at will, as exemplified by Theseus' ability to bring a bull out of the waters to slay his son, the king is still virile, therefore the culture is still dynamic therefore all is well. In ideographic terms, this myth means that the particular KING (not the culture) is still too powerful to overthrow; therefore, leave the culture alone for now. This is the interpretation of course, of those who would like to be that king and are waiting on the sidelines . . .

Bulls as Heroic Usurpers

In mythopoetic terms, when the culture becomes dormant, the world is on the throws of destructive chaos, which will turn things back to their original unmanifest state, which means for all things, their own destruction. When a king is too old or no longer virile, this is the message of the myth. Then, a young man, a hero usurper can come into the failing kingdom and infuse it with new vision and power, thus restoring a new cosmos from the chaos of the old order under the prior king. In ideographic terms, when the usurper is seen as a bull it means that he has surpassed in virility the past leader and therefore he deserves to be the king or "lead stud" of the kingdom.

Ox

Finally, we come to the ox, the castrated bull and his place in myth. In the mythopoetic interpretation, the ox represents a male who sacrifices one kind of knowledge (procreative) in order to achieve another kind of knowledge (imitative creation, such as an artisan or craftsman, scribe or priest might do. In women, they remain virgins in order to give up their procreative knowledge for this kind. See for example Athena.) They do this because although they are devoted to the goddess, they cannot create like she creates, but they choose to emulate creation of forms in the way a goddess (or woman) creates forms; thus, they perform the same act with a different medium, and with different results. Nevertheless, their goal is creation and they create IMAGINITIVELY. In ideographic language, the ox represents the worker (as in the agrarian model of the ox as plow mover) who will never have the power to overthrow the king, but will be there to advise the king, because he has a special kind of knowledge or skill that the king, since he is still virile does not have but nevertheless wants. In particular, these "oxen" of the king consist of the royal advisors and priests, who manipulate the political and economic affairs of the king, without having the (you know what's) to ever overtly attempt to overpower him.

These are the basic elements of the bull motif and its varieties. Returning back then to our question of the characters in the Oedipus myth cycle, we see that all that is really occurring in these myths are various accounts, both mythopoetic and ideographic of the roles of men. Originally, these myths were meant to instruct young men initiatically on the functions of adult malehood; later these same myths came to account for the various vocations of men in an all male world. In the former case, men were taught the method by which they could become immortal through a re-creation of the union of the beginning, by mating with woman and producing progeny. In the latter interpretation, the myths were used to show men the types of professions they could hope to possess in their future and what they could expect as reward for their efforts. Thus, the mythopoetic rendering of these myths were primarily religious and of spiritual consequence; the ideographic ones became secular and of material consequence. And the shift from one to the other was brought about by the group responsible for transporting this myth system across the known world through the medium of the written word. Thus the Phoenician traders and seafarers created perhaps the first case in history of the messenger killing the message, instead of the other way around.

Seers and Monsters and Masculine Knowledge

It is evident from the analysis in the previous section that the true meaning of these myth cycles lie somewhere in the realm of some kind of masculine initiation. Judging from the inordinate amount of emphasis on variations of masculine roles in the story line and the virtual uniformity of feminine prototypes, one is left with the reasonable conclusion that at the superficial level at least, these are stories about males, for males. The curious fact is that although this seems to be the case in the myth cycle of Oedipus, the original Indo-European source out of which these myths came, are NOT exclusively masculine oriented in interpretation, though many have concluded that they in fact are. The truth is that it is not the text of the Indo-European Rg Veda which has a masculine slant; it is the READING that has been done on the Rg Veda which produces ONLY the masculine slant. The reason why this occurred, particularly in the case of the Rg Veda (and the argument could even be extended to our Greek example) is because of the difficulty in transcribing literally, a philosophical system that bases its comprehension NOT on substantive things but on POWERS that exist without FORM. By the human faculty of ideographic substitution, horrendous renderings of these powers become "manifest" as the objects in which the powers were inherently manifested, eventually causing virtual havoc in the realm of ritual, not to mention interpretation, especially in cases where the interpretation moved away from native shores and becomes transported into a different culture. This is the fate of the hymns of the Rg Veda; they suffered first at the hands of ideographic priesthoods, who created mimetic substitution systems out of the "powers" by first translating these unformed elements into objects, for example, Indra became the Bull; Agni became the goat; the Asvins became horses; the goddess became a cow, etc. Then after that initial transference from the unmanifest power to the OBJECT possessing the power, they created elaborate sacrifices as a secondary substitution transference of the powers from the objects (for example, the bull) to themselves or the ones performing the rituals. This is the birth of ritual magic: "stealing" the forces of creation inherent in one object and transmuting them into the body of the sacrificer. All of this came about because the priesthood learned the power of the masculine knowledge of the word, and using the same act of transference between the written word and the object, they extended this form of understanding from the ritual object to the sacrificer. In the mythic accounts in the Oedipus cycle, we see numerous renderings of the ultimate results of this kind of sacrifice as the various kinds of monsters in the myths which the heroes slay, purportedly taking on then the "powers" of the slain beasts. So when Bellerophon slays the bull to Poseidon, he asks for the transference of the power of the bull, just as when Indra in the Rg Veda slays Visvarupa he retains the powers of both the masculine and feminine creation of forms. Indra also slays the serpent/dragon *Vrtr* (Sansk. "to bind"), in the Rg Veda; likewise SHOULD Oedipus slay the Sphinx (Gr. "to bind") in our myth; BUT HE DOESN'T. Perseus and Theseus and Bellerophon each slay their share of villains: Perseus has his own sea monster similar to Indra's *Vrtr* and Bellerophon has his Chimaera (a composite of animals all relating to masculine virility: goat, lion, snake); and Theseus has the Minotaur (another bull). But Oedipus never slew anything: he TALKED TO IT. Is this a clue to understanding the problem of this character? Has Oedipus become SO IDEOGRAPHIC that he doesn't need to DO anything at all, only talk? Yes, his world is no longer in the ritualistic realm at all; he's not concerned with taking on any powers of any animal by sacrifice, because he is BEYOND that: for his ONLY "power" is TALK itself.

We should also make a note here with reference to the old kings' attempts to kill their progeny. This occurs for the same reason as hero's kill monsters: so that the "powers" of the youthful male will be bestowed upon the "killer." (This is probably another reason, albeit perverse, why wars were/are popular among men, too, though the original "logic" may have become obscured in light of the overwhelming sophistication of methods of killing that have come to pass over the generations. Alas.) So, as we recall, Laius tried to sacrifice his son; Perseus' grandfather tried to kill his grandson, etc. The reasons why the heroes eventually killed their forebears then was that these men were not successful in their endeavors to kill the youths; thus they themselves perished. The only one who didn't kill an ancestor remember, was Bellerophon; BUT HE DID KILL a man before his exploits; thus the "trick" of taking over their "powers" still worked! IT'S THAT SIMPLE. AN IDEOGRAPHIC SUBSTITUTION of powers for powers and the world is in the throws of maniacal behavior. What we MUST remember is that this seemingly insane behavior is a DIRECT result of the technology of ideographic transference.

In a similar vein, in both the myths of the Oedipus cycle and in the Indo-European hymns, we are faced with the oracular utterances of the seers. Once again we become faced with TWO interpretations of what these characters mean in their stories. From a mythopoetic point of view, remember that the seers see in the realm of the "uncreated," that is, they see imaginatively, before manifested forms appear. In this manner, their vision is divine, for while they exist as mortals, they serve as an immortal presence in a temporal reality, thereby appearing as archetypal models for other humans to strive towards. They are the closest thing to divine will on the mortal plane of existence. So, how do the seers become ideographized? When the priests begin to "imitate" the visions of the seers through the use of hallucinogenics, which create the illusion, in the mind of the priest of divinity (because of the marvelous "forms" which appear). Since these drugs can be MADE by the priests' own hands, once again they USURP the POWER of DIVINE INSIGHT from the hands of the seers to themselves and to anyone else who'd like to pay to receive it. So, kekyion was the favored drink of the Eleusinian mysteries; Soma was the Rg Vedic substance of choice; even the Delphic priestesses were said to have become intoxicated by the fumes spewing forth from Apollo's holy temple in Delphi, and that these fumes were in fact, the SOURCE for their oracular vision. So, in myth Tiresias represents the old ways of the seer; his vision is INSIGHT (hence his blindness); the Delphic priestesses represent the new way of receiving divine revelation, from the OUTSIDE, by means of a hallucenogenic (a SUBSTITUTION system for knowing imaginitively). But does Oedipus take any hallucenogenic to understand the will of the gods? NO! What does he do? He Gouges out his eyes as if that were going to give him any kind of divine knowledge! He substitutes NOT the ACTIONS of the SEER in his own life, but THE VISIBLE IMAGE of the seer, so he blinds himself! Here's ANOTHER CASE OF OEDIPUS' MISTAKE OF INTERNALIZING THE HUMAN TECHNOLOGY OF SUBSTITUTION, of course, to his demise.

❁ ❁ ❁

The Heresy in the Myth

From our examination of the myths in the Oedipus cycle so far, we realize that nearly all of the members of the house of Zeus and Io are absolutely guilty of appropriating ideographic knowledge to the virtual exclusion of mythopoetic knowledge. We've seen it in the hero prototypes of Perseus, Theseus, Bellerophon, where their exploits can be simply understood in terms of an ideographic re-rendering of the mythopoetic myths of masculine initiation. We've seen it in the acts of Oedipus' ancestor Cadmus, who founded a city by the masculine method of "sowing the seeds of the dragon" in order to populate his city, instead of choosing the more "natural" route; i.e., one involving women. We've seen it in the virtual uniformity of female roles in the myths, which seem to serve only as poor placemarkers in the dynamic unfoldment of the storylines in this cycle. We've met the twin brothers who've populated the known ancient world, creating such monumental achievements as the introduction of patricide, matricide, genocide, suicide and the like in their wake. We've examined the origins of several names in the myth cycle to determine that they were indeed meant to be read as ideographic tokens of the acts they are to perform in the script. And we finally come now down to Oedipus himself. Why, for the last time, is HE the HERETIC? Because Oedipus is not even mimetic ideographic like his ancestors were; for he became the complete DISEMBODIED body of antiquity. While his forebears were on the throws of disembodiment, through the constant externalization first of powers of procreation, then later, powers of material creation, then powers of familial alliance, then powers of knowledge itself, until finally, we have Oedipus who has completely externalized even his own life AND DEATH! Oedipus is so outside of himself that, while he has royal origins, he questions them; while he kills a man on the road, he questions whether it could be his own father; while he procreates, he unknowingly does it with his own creator; while he believes himself a good king, he is the cause of his kingdom's demise; while he fancies himself a harsh defender of justice, he is the very person whom he should be judging and does not, for he cannot see within himself; while he rejects the divine will, he nevertheless continuously acts based upon it; while he hurls accusations at the inner vision of the seer, he imitates the EXTERNAL image of the seer when he discovers his sins; while he leaves Thebes for the good of the city, he curses his own lineage to exact its demise for what he says they'd done to HIM! Finally, when it comes time for him to close the chapter called his life, Oedipus even at the very end, waits for a voice from the outside to call him, whereupon he blindly follows and DISAPPEARS from the face of the earth, just like yesterday's news. His heresy is that he failed to learn mythopoetic life AND he failed to learn masculine ideographic life; in fact, he failed to learn human life, replacing it with a theory for it: logos.

✿ ✿ ✿

Summary and Conclusion

In the beginning of this chapter we outlined the development of a very special kind of "hero" myth paradigm; the paradigm that tells the story in MYTH of the birth of theorizing. In the myth cycle that followed that outline, we showed how the descendants of Zeus and Io brought forward this new "knowledge," by using the criteria of the myths themselves; namely, by showing how these "myths" differed in both content and style from previous myths of the hero, and offering a reason why this shift took place in the myth of this new hero: to account for the birth of ideographic theorizing IN the myth pattern itself. In other words, woven into the fabric of the story about the birth of theorizing are the actual CLUES to the historical development of theorizing. And as a result, these myths also clearly elucidate what we call the mind/mind split, culminating in the creation of the character of Oedipus.

The real root of the mind/mind split lies in a change of attitude among humans regarding the source of true knowledge. The trouble is that this change in attitude was sprung upon humans by a simple technological shift from primarily creating from the inside through the faculty of imagining, to exclusively creating from the outside using concepts as a medium upon which to develop theories about the world and ourselves. This habit of external thinking is reinforced by the habit of writing and the subsequent mental processes that the written word creates: induction, causality, magical thinking, substitutional transference first of pictograms for images, then symbols for images, then sounds for images, and then symbols for concepts until finally concepts for concepts without a human passageway back to the original source of the concepts. This leads to the need to create hypothetical origins which ought to be postulated and then later verified in the objective, outside, real world, making humans then, mere invariant observers of phenomena which exist external to themselves which at times can be beneficent to them, but at other times, malevolent. And this human, bound by the invariance of this model becomes a stagnant observer, always trying against all odds to find some meaning in this hopeless labyrinth called life.

The secondary consequence of this shift in mental habit occurs when someone proclaims that they've found the secret key to unlocking the mystery of life. Who are these people? The seers and the priests, who objectify further this human dilemma by creating yet another substitution system, called ritual, to aid the weary human in his search for origins. These holy men preach of "powers" inherent in "nature" and offer a way to transfer these malevolent powers into the human so that the human can harness them and become more "powerful" himself, so that he can overcome these forces that threaten to "overpower" him in this hostile world. So, ritual sacrifice is born, first of other humans, then of animal (direct object-object substitution), then the products of human hands such as agriculture or handiwork (indirect object-object substitution) then finally merely the recitation of words constitute the ritual ("I now pronounce you husband and wife." magical abrakadabra using the hidden "powers" behind the WORDS THEMSELVES entirely DEVOID of ANY REFERENT OBJECT IN THE WORLD WHATSOEVER!) Sooner or later, the particular substitution system strangles itself in its own conceptual loop, when the ritual itself has become so externalized that no one recalls its origins or usefulness anymore, yet THE HABIT REMAINS and we continue to go through the motions because it's easier to stay with the status quo than to go through the human exercise of reinventing the proverbial wheel, especially since all we can do, by habit is replace the old substitution system with a new one. Hence, the birth of the logos loop and the completion of the mind/mind split.

Oedipus is interesting to us because he represents the neural turning point in our human progress towards left brain exclusivity with regard to the realm of what constitutes knowledge, for it is with Oedipus that the mind/mind split became reality for the rest of us. Gone are the gods, the world of women, the world of images, of feeling, of fate and the conditions for human destiny interwoven in the cosmic plan. From the time of Oedipus on, we decide at each turn of the conceptual corner what to do, what not to do, and wallow in the agony of indecision and loneliness.

For better or worse, Oedipus chose for the rest of us what true knowledge was to become: "masculine knowledge," the knowledge of the left brain with its theoretical loops and mirages of infinity, virtually muting forever the inner voices of our other selves. And we can no longer shift away from this left brained dominance with ease, for it requires a Herculean effort to be creative in a world that encourages a repetition of this theory-making habit in every facet of life from making love to making dinner. And what's worse, we as a species can no longer even live without listening and authenticating the dictates of our own left brains, externalized to points of infinity, leaving us forever divided internally.

This mind/mind shift simultaneously increases our capacity for self deception. For one thing, we increase the mystery. We deceive ourselves into believing that we must continue deciding even if we are unsure of the result, or the reason to act, or even the theoretical outcome. We act now in the name of progress. And what is progress? It is the easy placebo that the left brain manufactures in order to keep itself active. It must invent its own gods, since it rejects the gods of the right brain. But can this left brain perform this task of invention? Creation? Ask anyone in the streets and they'll tell you not only that we've advanced as a civilization but specifically HOW we've advanced. But is this really true? What are the criteria for advancement? A more advanced, sophisticated substitution system? A "culture" which can function perfectly without the interference of "human error"? The mathematical formula that would for once and for all make humans obsolete?

Because of the mind/mind split, mother, father, family, life are no longer company, movement or emotional sensation. At the hands of Oedipus they have become cyphers, ghosts, objects, substitutes, the way a simulacrum substitutes an icon, or a concept substitutes an image, or an object breaks up the continuity of space and time. From this theoretical moment on, technology will substitute technology while Oedipus dreams of a human heaven without gods, which is delayed into the nebulous future for it can never come without bringing with its arrival the death of the species as Oedipus imagines it to be. The world of mythopoetic creation has given way to the world of ideographic representation: concept replaces image, theory replaces cosmos, the individual self replaces the will of the gods, the will of the nation replaces the individual will, the global market replaces national interests, writing replaces oral transmission, the book replaces the right brain as the source of knowledge, all is external to the self as it is to the book (for nothing happens IN the book), while everything continues to happen in the world. But the world is not the oral world of the right brain, but the exteriorized written world of the left brain; not a world of making images, but a world of images already created; not an immediate world of sensation, but a world at a distance, a world of mirrors, of pictures of electrical dots on a screen, a disembodied world, until world no longer feeds sensation and the human robot collapses.

Oedipus' heresy is simply that he not only split the brain into right and left, but that he rejected the right and lived only by the left: in doubt, in agony and in blindness. His death should be a warning to us all, for he disappeared without a human trace (without a human trace), except that we inherited from him the mind/mind split. We are all neurologically linked to one another, but it does not mean that we should continue to deceive ourselves just because we happen to have shared an eccentric ancestor.

CHAPTER FOUR

THE HISTORY OF THE HERESY

Introduction

In the last few chapters we have outlined the nature of what we have termed the mind/mind split from the perspective of the literary traces apparent in a variety of myth cycles. Now we shall turn to historical evidence in an attempt to link specifically the details enumerated in the myths with the material artifacts of their corresponding cultures to further define the necessity of viewing cultural history, not with an ethnocentric bias but rather through the criteria set forth in the biocultural paradigm. For as we shall see, "cultures," especially ancient ones, are all defined solely by a set of criteria which don't define "culture" per say, but only seem to delineate the permutations of ONE single biocultural bias that imposed a single set of criteria which then spread all over the known world, the remains of which constitute for US, the "signs" of "culture." In fact, they are merely technologies introduced by a single group and incorporated into the indigenous groups' own technologies, sometimes causing, as a result, the demise of that indigenous group. This is what we shall refer to in the future as cultural imperialism, and this imperialism began in antiquity with the ancient Phoenicians.

❁ ❁ ❁

So Who Were the Phoenicians?

If ever a group were severely underrated, grotesquely underrated, it is the ancient Phoenicians. An average history book usually covers their entire civilization in fewer paragraphs than you have just read in this chapter, while devoting scores of pages to their contemporary civilizations: Sumer, Egypt, Minoan Crete, even the Indus Valley Civilization. Yet, what everyone seems to forget is that all of these other great civilizations owe a majority of their heritage to none other than the Phoenicians; indeed, in some of the above cases, their civilizations are in fact genuinely and actually Phoenician. Yet to read about this mysterious culture called the Phoenicians proper, one is awed by their apparent lack of sophistication in comparison to these other great civilizations: their only true claims to fame for most historians appear to be the development of the alphabet, and the famous purple dye that they produced from the murex sea snails. According to traditional historical accounts, this group was also known for its great seaworthiness as well as its ability to mine and use all types of metals, and for the unflattering distinction of having been a rather rough and generally unfavorable type, as the following quotes indicate:

From the Bible[1]:

The Phoenicians were descended from Ham, the cursed son of Noah; specifically from his son, Canaan (hence their ancient name, the Canaanites.) Noah reflected the sentiment that the ancient Hebrews had on this group in Genesis 9:26, "Cursed be Canaan; a servant of servants shall he be unto his brethren." Note that this curse took place AFTER the proverbial flood, a fact that shall shortly become significant.

From the Greek Historian, Herodotus (5th c. B.C.E.):

Regarding the grounds for the eternal feud between Greeks and barbarians, Herodotus reports in Book One of the Histories, " . . . the Phoenicians began the quarrel. This people who formerly had dwelt on the Persian Gulf, having migrated to the Mediterranean and settled in the places where they now inhabit, began at once . . . to adventure on long voyages, freighting their vessels with wares from Egypt and also Assyria. They once landed in Argos . . . where they exposed their wares and traded with the natives for five or six days . . . (W)omen were standing by the stern of the ship intent upon their purchases, when the Phoenicians, with a general shout, rushed upon them. The greater part made their escape, but some were seized and carried off; Io, herself being one of the captive. The Phoenicians put the women on board and sailed off for Egypt. And this is how Io truly arrived into Egypt . . . "

From Plutarch (1st c. A.D):

"They are a people full of bitterness and surly, submissive to rulers, tyrannical to those they rule, abject in fear, fierce when provoked, unshakable in resolve, and so strict as to dislike all humor and kindness."

From Pomponius Mela (1st c. A.D.):

"The Phoenicians were a clever people, who prospered in war and peace. They excelled in writing and in Literature and in other arts, in seamanship, in naval warfare and in ruling an empire."

The above references which describe a group known as "Phoenician" or "Canaanite," mark its common affiliation from about the eighth century B.C.E. But what about the earlier "culture" of the Phoenicians? Should one assume that prior to the eighth century B.C.E., no such people existed? The biblical tradition places the sons of Noah as living in approximately the second millennium, B.C.E., which would place the existence of the Canaanites at about the same time. The myth of Cadmus the Phoenician's founding of Boeotian Thebes that we introduced in the last chapter, according to the archaeological records of the site, is also dated at about the second millennium, yet scholars are at a loss about this fact since the Phoenician alphabet was said not to have been introduced into Greece until the ninth/eighth century B.C.E. An even further problem arises as to the origin of this "culture" when we realize just what it takes to MAKE an alphabetic system. Surely, no one would even attempt to conjecture that a heretofore unknown group SUDDENLY appears with no real "culture" except a mature alphabetic system! It seems rather the case that something is surely wrong with our understanding of exactly who these Phoenicians truly were. And it also seems that the key to unravelling this mystery lies with the key to unravelling another mystery of culture: the mystery of the universal flood myth. For perhaps the answer to what happened to this vast seafaring culture known as the Phoenicians between the second millennium B.C.E. and the ninth century B.C.E. can be easily explained by the series of cataclysmic disasters that rocked the ancient world during that historical period; not the least of which included the one in which a large portion of the ancient island of Thera in the Mediterranean literally disappeared. And perhaps the tattered remnants of those Phoenician people, when they settled in the coastal regions across the ancient Mediterranean, brought with them, after the devastation, three things: their various human technologies, their account of the great flood as told in the myths AND as a consequence of both of the above, the mind/mind split.

❁ ❁ ❁

The Textual Evidence of The Phoenician History

Prior to addressing the myth of the flood and its relevance to our understanding of Phoenician history, it is crucial to note one very important written source on Phoenician history, translated into Greek by Philo of Byblos, during the Hellenistic period and preserved for us in fragmented form, by Eusebius, the fourth century Church Father.[2]

The original document was allegedly written by Sanchuniathon, a Phoenician whose dates have ranged from the second millennium, B.C. to the second century, A.D. (contemporary with Philo.), with the preponderance of the evidence leaning towards the earlier date (because of the similarity between the cosmology in Philo's account of the Phoenician History and the recently discovered Ugaritic texts, which have been dated to that time.) In any event, the fragments preserved in Eusebius' history display such uncanny similarity to the remnants of the "so called" logos myth cycles that we have been heretofore discussing, that a few selections are worth noting here.

On the Phoenician Idea of Deity:

> The most ancient of the barbarians, and especially Phoenicians and Egyptians, from whom the rest of mankind received their traditions, considered as greatest gods those men who had made discoveries valuable for life's necessities or those who had in some way benefitted their nations. Since they considered these men as benefactors and sources of many blessings, they worshipped them as gods even after they had died. They built temples and even dedicated steles and staves in their name. The Phoenicians paid great honor to them and instituted magnificent feasts for them . . . Among things in nature they acknowledged as gods only the sun, the moon, the other planets, the elements and their combinations, so for them, some gods were mortal and some immortal.[3]

On the History of Kronos and his Offspring

According to Philo of Byblos, Kronos was not, as Hesiod of Boeotia wrote in his *Theogony*, a Greek god, but a Phoenician god, meaning a man who had done such good works in his life that the people immortalized him through veneration. His father, Ouranos was, likewise a man and his mother, Ge was a woman. Ge and Ouranos' parents were called Elioun (father) and Beirut (mother), again two PEOPLE. When Elioun died, his children Ouranos and Ge worshipped him by offering funeral libations to him.

When Ouranos and Ge married they produced four children: ELOS (Kronos); BAETYLOS; DAGON ("Philistine" grain god) and ATLAS. The history relates that of all the children, Kronos became the most powerful one and by overthrowing his father, the leader of all his people. Atlas was "cast down into the depths of the earth" because Kronos was suspicious of him. (Nothing is said in the History of the fate of the other two children.)

Kronos' children were:

Persephone (who died)

Athena (to whom he gave the land of Attica)

Sadidos (whom he killed)

Apollo by Peraea

Zeus Belos by Peraea

Kronos ll by Peraea

7 Artemids (Titanids) by Astarte

Two sons -Desire and Love by Astarte

Two unnamed daughters by Dione

Seven Sons by Rhea, including Muth, who died and was worshipped as Pluto

He also had, by the lineage of Zeus Belos, a grandson, Nereus; a great grandson, Pontus and twin great, great grandsons, Sidon and Poseidon.

Kronos was the "god" of Byblos and through his affiliation with the "gods" of Tyre, he became the supreme ruler of the Phoenician world. It should be noted that his "allies" from Tyre consisted of such figures as Thoth (remember the Egyptian "god" of writing? Same guy.); the Dioscuri; the seven

Kabeiri; the Korybantes; and Askleipious. These allies were given the name ELOIM, because they were friends of Kronos (Elos.) The lineage from Byblos were called Kronians; and presumably, those from the lineage of the lost brother Atlas were called, Atlanteans.

On the Theme of Castration and the Phoenician History

Remember when we were perplexed in the last chapter about the role of castration in the myth cycle of Oedipus? Well, be perplexed no more, for the answer to why castration was a sign of being in the "family" or one of the god Kronos' chosen people is written in the Phoenician History. The story is as follows:

> At the occurrence of a fatal plague (A FATAL PLAGUE!), Kronos immolated his only son to his father Ouranos, and circumcised himself, FORCING THE ALLIES WHO WERE WITH HIM TO DO THE SAME. (Emphasis added.)[4]

Presumably, the reason this was done was to appease the wrath of his father Ouranos, for, as the history is told, Kronos, in overthrowing his father, had Ouranos castrated:

> In the thirty-second year of his own assumption of royal authority Kronos, trapped his father Ouranos in a certain inland place. He overpowered and castrated him near springs and rivers. There Ouranos was made an object of worship and he breathed his last and the blood from his genitals dripped into the springs and the rivers' waters.[5]

Is this not a case of the sacred king being overthrown by a younger version?

Is this not an historical account of the fate of the leader when the ideals of ideographic knowledge take over, causing humans to create substitution systems based on their own logos understanding of creation? Haven't you heard this story before? Of course you have, but in the last chapter, we referred to it as a myth; now it is HISTORY.

So what of the logos myth cycles? Is it possible that they are not at all mythic? Is it possible that they are indeed the first example of ideographic writing: HISTORY? And is it possible that the transmission of this "history" was actually "Mythologized" by those biocultures who heard the story but were unfamiliar with the ideographic interpretations, thus rendering them, in such places as Greece, as myth? It seems that this is a VERY LIKELY situation.

The next section deals with yet another perhaps historical account that has become mythologized in the same way, though it probably did have its origins somewhere in the Phoenician History. That event is the so-called universal flood myth. Perhaps by reconstructing the true Phoenician history out of the mythopoetic mire, we can truly begin to understand the full import of this logos-bearing group.

✿ ✿ ✿

The Great Flood: In History and in Myth

If you were educated in the Judeo-Christian tradition, surely you have read or been told about the great flood of antiquity which wiped out the entire human race, save Noah and his family.[6] I recall as a child imagining that huge ark carrying two of every kind of animal floating for days upon days on the vast earth with no land in sight; in fact, that story in particular always terrified me because I wondered if it could ever really happen again, and would I be one of the people who would drown. Needless to say, I'd cringe at the sight of any substantial rain fall and, probably like many other children, wonder if this time, we'd finally done something so terrible that the flood was surely upon us again!

What's peculiarly interesting about the flood myth in general is that it is so universally known; of course, with different names of the survivors and naturally, different gods responsible for it. This has led mythographers and historians alike to attempt to discover whether there was in fact such a disaster in antiquity as to warrant the consistent retelling of this story. It turns out that there was: the explosion of the volcano at Thera in the Mediterranean.[7] What is particularly interesting and important for us is that this event and the ensuing natural disasters following the initial event seems to have destroyed, overnight, perhaps the most powerful and advanced civilization in the then known world: the Phoenician civilization.

From Kalliste to Thera: The History of the Disaster

Some time during the seventeenth/sixteenth century B.C.E., there was a cataclysmic event that shocked the entire ancient world, literally. The tiny round island of Thera exploded due to the eruption of a volcano situated in the center of the island. This volcano, it is said, had been dormant for some ten thousand years prior to this cataclysmic outburst, though after its initial explosion, it continued its sporadic activity throughout the centuries following.

According to geological research, the strength of this particular explosion exceeded the famous Krakatoa eruption of 1883 by at least six times; Mt St Helen's was child's play compared to Thera. So, what EXACTLY occurred on that infamous day on Thera?

1. The volcano erupted, sending the entire midsection of the island of Thera into the water: it literally sank.
2. The volcanic ash, death clouds and dust spread through the atmosphere landing as far away as Southern Turkey, where it spread over a foot of ash. If the people didn't die from the heat, they would have been instantly suffocated by the noxious gases. All animal and insect life would have been instantly killed as the black clouds passed through the region.
3. Egypt received less than an inch of dust; nevertheless, this small amount was enough to cause the sun to be blacked out for several days, and the harvest of the following year to be virtually non-existent.
4. Two inches of ash fell on Crete and in addition, tidal waves wreaked havoc throughout the western Aegean. An eight hundred foot tsunami rushed through southern Turkey.
5. Isolated earthquakes rocked the Mediterranean region, due to the shift in the tectonic plates.
6. The thinning ash spread across Syria, Iran and as far as Asia, causing storms of blue lightning and dramatic climate shifts.

And the above events are just the basics.[8]

Now imagine for a moment, a culture whose vitality depended upon its seafaring abilities; whose "homeland" consisted of several island based headquarters and coastal regions; whose economy depended upon trade. What effect, do you suppose, would such an event as the eruption of Thera have had on that culture? Do you imagine that such a culture would have had even the remotest chance for survival as a defined group immediately following this event? Is it not obvious that whatever survivors, would have been thoroughly terrorized by living through such an event, and would certainly have been changed as a result of this event? Don't you think that the human psyche would have retained the memory of this cataclysm and would have certainly altered its prior perceptions about the meaning of life? (Ask any survivor of a natural disaster about their lives before the disaster and after the disaster, and you will see what I mean here.) What do you think the majority of the survivors of this culture would have had to do to continue their lives after life as they knew it was virtually destroyed? And lastly, how could any human being NOT consider that perhaps there was a link between these cataclysmic events and divine will. In other words, wouldn't it have had to cross their minds that perhaps though their god/gods caused the disaster they were somehow responsible for it?

The "historical myth" of the Flood and the Phoenicians

The story of Noah notwithstanding, if ever there were an account of a lost civilization that piqued the imaginations of humans throughout the centuries, it would surely be Plato's myth of Atlantis. Though the veracity of the myth is considered dubious at best by modern scholars, there are more than a few pieces of information contained within that story that lend considerable support to the thesis that in fact, there was a great civilization that controlled the Mediterranean at some time in Greece's distant past (though perhaps not as distant as Plato would have had us believe) AND that from the description of this culture as he describes it in the *Timaeus* and the *Critias*, it could have been none other than the culture which we presently identify as the Phoenicians. In fact, it has already been very concretely postulated that Thera was the island that Plato referred to in these dialogues as Atlantis.[9] Certain inconsistencies in the text though cause many to disregard this thesis, continuing rather to view this work as just another piece of didactic fiction. Yet Plato took great pains to substantiate the story, as did others in antiquity, concluding that the "myth" is actually history.

For our purposes this myth is of interest because if taken as history, it explains the early onslaught of ideographic thinking in the middle second millennium B.C.E. and the apparent "dark ages" between the second millennium and the resurgence of ideography (incidentally, by the Phoenicians) in the ninth century B.C.E. It explains the preponderance of Phoenician style iconography (bulls for example and intricate artistic metal work, and fabulous constructions using unique forms of masonry) throughout the Mediterranean and Levant, particularly in the coastal regions; it explains the universality of certain myth cycles among the Egyptian, Greek and Semitic groups, and most importantly, it supports the mythic account of the migrations of peoples throughout the ancient world. Now all this can be easily understood and perhaps even found verifiable assuming one element: that the reader suspend his/her notion of ancient history as it had been taught for generations and entertain, for a few moments, the following presumptions:

a. That what has been termed the Minoan civilization is in fact a branch of the ancient Phoenician empire.

b. That the great Mycenean sites of antiquity are likewise branches of this Phoenician alliance.

c. That the Philistines of the Old Testament were in fact remnants of the ancient Phoenician alliance.

d. That the Egyptian cities of Memphis and Thebes were predominantly Phoenician and that for a time, all of Egypt was in fact RULED by Phoenicians.

e. That the mysterious "sea peoples" of antiquity were of Phoenician origin.

f. That Harrapa and Mohenjo Daro in the Indus Valley were either Phoenician trading outposts or actual Phoenician settlements.

g. That Greek is in fact derived from the Phoenician SEMITIC origin; having passed from Linear A (the most obviously Semitic) to its modern form in successive stages.

h. That the story of the Indo-European/Indo-Iranian/Aryan invaders is the real fiction; in other words THIS NEVER OCCURRED. Rather what happened was a universal acculturation through the technological advances of the Phoenician peoples on the rest of the world, thereby causing the cultural shifts usually attributed to these Indo-Europeans (such as patriarchy, sky god worship, and the disintegration of society.)

i. Finally, that the HISTORY of the above events is already blatantly

> evident within the existing knowledge that is available through our archaeological and literary sources; the only problem we have putting it all together is that we, as a culture are unable to "see" it anymore because we've so compartmentalized knowledge that we can no longer see the forest for the proverbial trees.

Beginning then with Plato's myth of Atlantis as a backdrop, let us reconstruct our human heritage, not through the criteria of ethocultural affinity but rather biocultural affinity, and thereby uncover for once and for all our true origins as a species.

The elements in the Atlantis story that we shall focus on are the following ones:

1. The geographic parameters of the Atlantean alliance.
2. The religion and political organization of the Atlanteans.
3. The chief products of the Atlanteans.
4. The layout of the city and the building plan of the surrounding areas.
5. MOST IMPORTANTLY, Plato's reasons for even bothering to write about Atlantis.

According to Plato's account (which is the only true literary account of the existence of Atlantis that is extant) the territory of Atlantis spread from the Pillars of Heracles (at Gibraltar) to Tyrrhenia and included Egypt and Libya and most of the islands within the Mediterranean proper. From the description in Plato it is clear that the culture of Atlantis was NOT confined to a single place, but a was rather a confederacy consisting of several geographic regions, though joined together under a single leadership, that being the on island of Atlantis proper, which Plato says was situated in front of the Pillars of Hercules. PLATO IS VERY CLEAR IN HIS DESCRIPTION THAT THE KINGDOM OF ATLANTIS SPANNED MANY GEOGRAPHIC AREAS AND WAS NOT MERELY THIS ISLAND.

Comparing the geographic areas of legendary Atlantis with what we factually know about the Phoenician settlements in antiquity, we find that the territories are in fact, identical, if we include the "Minoan" culture among the Phoenician settlements (and there are more than just a few reasons why we should do that, enumerated below.)

Not surprisingly, Poseidon was the god of the Atlanteans, and that god populated the race by fathering five sets of male twins who became the ten leaders of the region. Naturally then, the bull was the animal of religious preference in Atlantis; in fact, bulls were so highly esteemed that they were permitted to roam freely within the sanctuary of Poseidon. According to Plato, the ritual in which these ten kings received their autonomy from Poseidon himself is described as follows:

> Each of the ten kings was, in his own territory and government supreme . . . but their authority over and intercourse with one another was regulated by the commands of Poseidon, as they were informed by the law and by an inscription left by the earliest kings on a column of orichalch preserved in the sanctuary of Poseidon in the center of the island . . . They assembled there on alternate intervals of four and five years . . . and deliberated on their common affairs, made inquiry whether

> any of them were transgressing the law, and pronounced judgement . . . They exchanged pledges in this fashion. (Each prayed to Poseidon to aid them in the capture of the bull of their preference. Then they chased it with wooden clubs, for iron was not permitted. Then they slew whatsoever bull they took over the column whereon were pronounced the laws, wetting the inscription with blood.) Now there was written on the column, besides the laws, an oath calling down grievous curses on the disobedient. So when they had offered sacrifice after their own ritual and were devoting all the bull's members, they would mingle a bowl of wine, casting in one clot of the blood for each man; the rest of the blood they cast into the fire, first cleansing the column. Then they drew the wine from the bowl in golden beakers, made a libation over the fire, and swore an oath that they would give judgement according to the laws upon the column, would chastise . . . transgressors, and transgress none of these ordinances . . . neither giving nor obeying commandment save according to the laws of their father. When each had taken the vow for himself and his house after him, he drank and dedicated his beaker in the god's sanctuary . . . When dark fell, . . . all vested themselves in fair robes of deep blue . . . thus they gave and received judgement . . . When morning came, they wrote the judgements on a plate of gold and dedicated it and the robes for a memorial . . . [10]

There are more than a few coincidences in the above account of the Atlanteans which would warrant investigation as to their possible Phoenician origins. We have already seen in the last chapter that Poseidon was certainly the chief deity in myth at least, of all territories presumed to have been founded by the Phoenicians. Secondly, we are reminded once again of the appearance of male twins in the Phoenician myth cycle of the house of Io. From all accounts, it seems that bulls were in fact the animal chosen to represent the Phoenician god Baal, whose name means simply, king or lord;

hence it is possible that the origin of this "god" is none other than a late misreading of a royal icon, and not a name of any divinity, per se. The slaying of the bull recounts the bull sacrifice specifically in the Phoenician stories about Bellerophon, or even Theseus, that we recall from the previous chapter. Further, the historical Phoenicians were known for not using iron tools within sacred precincts (see the story of the building of Solomon's temple in the Old Testament); they were likewise known for building temples containing two columns upon which were recorded their laws; and finally, the " fair robes of deep blue" which they wore on Atlantis is an all but give away that the group was in fact Phoenician; for as we have said earlier, if there is anything that is universally known about the Phoenicians it is that they were famous for their purple dye!

Plato recorded that the Atlanteans possessed enormous wealth, more than any other royal line had ever had, because of their external revenue from the empire at large and also because of their natural resources on the island itself which were:

-all types of mining including orichalch, which was a substance more valuable than any other but gold, but is now unknown.

-substantial forests which yielded a generous supply of timbers.

-a developed method for domesticating animals, including elephants.

-varied agricultural products including roots, stalks, canes, gums dry fruits and oil.

He reported that there was constant commerce and trade in the region and that their harbors were always filled with triremes in excellent working order. The above account could have been used equally well to describe the products of the ancient Phoenicians. One has only to recall the famous timbers of the Lebanon forest used in the construction of Solomon's temple to realize that it was the Phoenicians who certainly cornered the ancient market with that trade. Their supremacy in the mining trade is likewise well documented, as is their knowledge of animal husbandry. But primarily, the Phoenicians were known as the merchants par excellence; curiously, another facet of Atlantean life which they share. Finally, the physical structure of the cities of the historical Phoenicians bears an uncanny resemblance to the layout of the city of Atlantis as described by Plato. According to Plato's account, the island of Atlantis was constructed as follows:

> They began on the seaside by cutting a canal to the outermost ring . . . ; the ring could now be entered from the sea by this canal like a port, as the opening they made would admit the largest of vessels. Further, at these bridges they made openings in the rings of land which separated those of water, just sufficient to admit the passage of a single trireme, and covered the openings so that the voyage through them became subterranean, for the banks of the rings of earth were considerably elevated above sea level . . . The islet on which the palace stood . .was enclosed with the rings and bridge, surrounded completely by a stone wall, building towers and gates on the bridges at either end of each passage for the sea water.[11]

The Levant coast, homeland of the Canaanites, possessed a unique coastline: many small bays adjoining headlands where they could easily defend themselves against landward attack while simultaneously enjoying good anchorage on either side for their numerous ships. Whenever they settled a colony, the Phoenicians likewise sought this type of physical layout; usually searching out a coastal region with a nearby peninsula or else island (which would serve as the barrier region). They purportedly occasionally even constructed artificial land masses which linked these islands to the coastline when there was no natural point of entry. Thus, in their homeland, the small island of Aradus held the power over the nearby regions of Marathus and Simyra; Cadiz ruled the geographic regions near Spain; Valletta controlled Malta; Bizerta in Tunisia; Cagliari in Sardinia; Motya in Sicily; Mogador in Morocco.[12] If we should extend this coastal theory a bit further, we realize that Cyprus likewise protected the Levant coast; the island of Thera (Santorini) protected Crete; and that the famous circular port of Carthage, could be none other than an artificial construction based on the same architectural principles. Of course all of these barrier regions were further protected by fortification walls constructed in almost a cyclopean masonry style.

Thus, perhaps Plato's account of Atlantis, which describes a culture whose boundaries extended westward to the Pillars of Heracles was in fact the Phoenicians, and the description of the island of Atlantis proper, was a generic description of the types of settlements made by this culture. Or perhaps, as many have now come to believe, Atlantis was in fact, Minoan Crete, the island of Santorini being the city which was destroyed overnight. Or maybe (most probably) the group whom we identify as Minoans, were in fact yet another settlement of Phoenicians and therefore the entire myth seems to make perfect sense: a huge culture, whose boundaries extend as far west as the Pillars of

Heracles was destroyed overnight, by the sinking of (one of its many) main citadels, Santorini. The subsequent demise of the entire EMPIRE would be a consistent consequence, because of the secondary effects of the eruption/sinking of Thera, ESPECIALLY TO A CULTURE WHOSE MAIN SETTLEMENTS WERE: COASTAL, ISLAND and/or ARTIFICIALLY CONSTRUCTED, whose chief means of power lay in its naval capacities and whose chief economy was trade. Add to these factors, the final issue, Plato's true reason for writing about Atlantis, and the entire mystery becomes blatantly evident. The Phoenicians, the traders/writers/oligarchs/barbarians of antiquity were threatening in Plato's OWN time to once again rule Hellas and he was violently opposed to this political maneuver. So, what was Plato so afraid of? After all, we say, technological achievements such as alphabetic writing, advanced architectural design, sophisticated urban developments and steadfast systems of jurisprudence all lead to productive, stable states. And who could argue with this?

The problem is now, and was then, that no one was able to argue because these things became the sole CRITERIA for cultures' to define their own progress. By contrast, what was/is termed primitive, were/are the basic human technologies that produced such "unnecessary" qualities as art, music, literature and philosophy, right? Wrong. For Plato in his era, the Atlantis story was introduced by analogy, to remind the intelligentsia of the time of the mercantile invaders and their MINDSET, in an attempt to rid his region from the INTERNAL onslaught of barbarian technology — writing, commerce, war and political tyranny — to the exclusion of internal human creation. In short, his dialogues were meant to perpetuate the need for all the human faculties, especially the acts of the mythopoetic biocultures. Unfortunately, even his works have come under the scrutiny of the mind/mind split, thereby all but smothering this great visionary's human project.

And now we come to our own inheritance. What of our history and the effect of the mind/mind split? A simple walk through our human history tracing the origins of so-called "civilization" reveals uncanny insights as to the role that the "primitive" acts of humanity have played in the shaping of modernity. Actually, it appears that these "primitive" insights form the CRUX of our human legacy, DESPITE the intrusion of the logos bioculture and its insistence on ideographic superiority. It is certainly time that we, as a global culture, composed of varying though similar biocultures, accepted THIS history as our own, and no other.

❁ ❁ ❁

Aspects of "Civilization" and the Mind/Mind Split

Ask any social scientist to define what makes a "civilization" and you will usually receive the following rudimentary list:

-an established political system or evidence of urbanization

-an established socio-economic base

-a distinct system or systems of worship

-a means for transmitting knowledge (some would go as far as requiring a written means of transmitting knowledge)

-a body of literature or history defining the group as self-identifying or evidence of familial genealogy

For purposes of this discussion, let us take these elements at face value and attempt to describe, not the historical developments, but the biocultural developments of these elements in an effort to uncover exactly HOW the mind/mind split forever has altered our perception EVEN of our own, global "civilization."

❁ ❁ ❁

Political Systems and the Mind/Mind Split

If one were to begin the story of political systems in history, one would be faced with the first absurd realization: that political systems were first based on blood lineage, as traced through the maternal lines (as the male role in procreation was not immediately evident). (Read here, maia bioculture.) The earliest civilizations then were matrilineal successions with the masculine role forming the less important position because of its lack of capacity to regenerate. Once the masculine role in creation was defined, then cultures', still based upon matrilineal succession, appropriated male counterparts to aid in the process of regeneration and fertility. (Read here, mythos bioculture.) The problem with this system, if you happened to be a male, especially the consort of the female in charge, was that once your fertility wore out, or there was a plague or famine in the land, or the female found a new consort, your job was literally over. What became of you, however, depended upon the BIOCULTURAL interpretation of the society in which you lived. In maia/ mythos biocultures, you left the role of householder and prepared yourself (internally) for your spiritual future (reincarnation, for example.) In other words, you practiced the art of dismembering your biographical self (the one you built up by your attachments to the female) and readied yourself for re-creation. ENTER, THE MIMETIC BIOCULTURES.

The masculine solution, if the culture is right brained mimetic is slightly different. Here, the consort when through with his duties is raised up to the position of deity itself, to match the position of deity that his female companion already possesses (as she is able to create life). The problem is that HE MUST DIE TO BE MADE IMMORTAL. The solution? In right brain mimesis, the "hero" is sent to war or some other dangerous assignment in order to earn his right of divinity. But in left brain mimesis, the males simply create a human substitution system for themselves and sacrifice him. In short, they find a substitute victim: a scapegoat.[13] First, the substitute was another human; later a sacrificial animal; sometimes their own offspring. The shift in left brain mimesis to substitute victim brings with it another phenomenon that developed simultaneously with this: the belief in societal guilt and atonement. For, at these sacrificial rituals, all the sins of the community were symbolically placed upon the victim so that, upon its death and resurrection, the community maintains its pristine character. ENTER THE LOGOS BIOCULTURE. With the shift to substitution systems, the masculine role of superiority gains political momentum. Now, the kings are the benevolent gods on earth whose power can be evidenced in the many material developments the communities (which are now stable) can enjoy. Now lineage can be traced through the paternal line because the human technology of substitution systems gives way to artificial methods of defining knowledge. Now creation of life itself is no longer the sole criterion for political systems, as a whole range of other manifestations of (masculine) creativity can be established and appropriated. And finally, we come to the age of the divine rights of kings, of the noble houses, of the political machine.

❁ ❁ ❁

Socio-Economic Systems and the Mind/Mind Split

The fulfillment of the human needs of food, clothing and shelter form the crux of any socio-economic system. These items were easily taken care of in agrarian societies, or those societies that practiced domestication of animals, through the simple distribution of labor between the sexes. But how could a society flourish without these essentials? What if their inhabited space was not responsive to agriculture or animal husbandry? The response is a sort of twofold/singlefold one: steal the items from those who do have them, or devise a system of trade. Wars, of course were the result of the former solution. For the latter solution, again we must refer to our biocultural history, for in order for there to be a conceptual understanding of trade, cultures' must first possess an understanding of quantification, differentiation and substitution systems; in short, the society must be at least of a biocultural mimetic type. Simple barter, or item for item trade can of course be evident in right brain mimetic systems. Commerce through currency or artificial substitutions, though necessitates left brained biocultural traits. But the most heinous of all biocultures, when it comes to this category is the logos one, for this bioculture appropriates its societal needs by means of LAW. That is, it sets up an artificial substitution system not based on objects of trade, but on WORDS THEMSELVES which the system holds to be self evident and on the authority of the WORDS themselves, literally plunders the societies of the other biocultures in the name of that

Almighty Law.

❁ ❁ ❁

Religious Systems and the Mind/Mind Split

The earliest form of worship was clearly goddess worship because, quite simply, women were perceived to be the source of regeneration and thus immortality. This understanding aside for a moment then, how did we ever devise such elaborate systems of religious worship as exist today, with such varied and contradictory precepts? Enter, the mind/mind split again.

There are two basic understandings of the origin of deity in early society, goddess worship and celestial worship; but both curiously enough, share a common beginning. That beginning is the moon. Matriarchal societies first posited the cyclical pattern of nature from observations linking the changing seasons, the female menstrual cycle and the planting and harvesting of agriculture. This "wisdom" of the knowledge about the regularity of the lunar/solar cycles provided enough information to establish communities and advance the stability of the fledgling race known as humans. Since women, and not men, were privy to this cycle internally, they were naturally perceived as being closer to the plan of creation than were males. By their affinity to creation, it was believed that they, in copying the geometric archetypes as found in nature, were able to create mortal forms. Of course, geometry was invented by these cultures, and was defined by them as the matrix of forces which must appear for manifestation to occur. And since manifestation did occur through (in) women, they were held in esteem as the holders of this immortal power (the power of form-making through manipulation of geometric matrices.) Hence the preponderence of goddess worship as well as the birth of matrilineal society. In effect, the POWER of the moon was perceived as being somehow linked to the female; hence, lunar worship in association with goddess worship. Later, with the acknowledgement of the dual role of male/female creation, the sun, whose heat was responsible for the powers of agricultural growth through the replication of the same geometric forms, became linked to the male, to serve as the lunar counterpart to the woman. (Occasionally, this duality, based upon the same principles was seen as the sky/earth duality instead of the sun/moon one, with the same results.) Polytheistic religious systems developed when other human powers were established in archetypal form (in a way that they could be learned and replicated by each member of society); hence the veneration of the gods as ideal archetypes. Enter the Mimetic biocultures.

In right brained mimetic systems, where POWERS of creation are not posited but OBJECTS of creation are, the sun and moon became actual personifications of deities to be worshipped AS OBJECTS. Naturally then, a substitution system developed an "as above, so below" tenet which yielded the solar male/lunar female iconography in early systems of worship. Of course THESE SO CALLED DEITIES WERE OF TWO KINDS: the immortal unchanging ones (called later the geometric archetype, incorrectly of course) in the sky AND THEIR COUNTERPARTS ON THE EARTH: the "male and female" leaders of the community. Following this then, any hero who became immortalized would be apotheosized into a celestial being, usually a star or a constellation, which retained some geometric form, though its substance was one with the immortal deity, the sun. These are the origins of the myths of constellations and the like.

In left brained mimetic systems, though, religious practice went one step further. No longer was the physical person the object of worship but THE WORDS THAT HE /SHE SPOKE became likewise the immortal vestiges of wisdom, as of course an extension of themselves. Now, in addition to the creation of deities on earth, another entire class of individuals had to be created in order to interpret the wills of these two classes of deities: the priesthood. Temple/Court astrologers, diviners, haruspices, and others interpreted the celestial gods' wills; dream interpreters and scribes interpreted the terrestrial gods' wills. Now geometry, the study of the immortal archetypes of natural forms became gematria, the study of the numerical/magical relationships between alphabetic letters, numerals and symbols inherent in any word of any terrestrial deities utterances. And of course with the mind/ mind split, the WORD itself became the omniscient, omnipotent deity, used to interpret IN PLACE OF ALL THE WILLS OF ALL THE OTHER DEITIES.

❁ ❁ ❁

The Transmission of Knowledge and the Mind/Mind Split

The transmission of knowledge and of course what constitutes knowledge is really the subject of this book. We have seen how, with the advent of alphabetic systems and its ensuing mindset, knowledge has been transferred from being apprehended from the "inside" to the "outside"; from our neurological sense centers being the ultimate decision makers in our choice of human behavior to the "objective laws" determining our behavior patterns, with our only choices being to accept or reject them, with the ensuing results. Why is this so bad? Because accepting as knowledge only that which can be verified by the criteria of only one single aspect of our many neurobiological ways of knowing is tantamount to censorship of the worst kind: biological! How can a mythopoetic bioculture justify ideographically when the means of apprehension share no correlatives? Further, WHY must the mythopoetic bow to the ideographic and not the other way around? (After all, the brains of the mythopoetic biocultures, in fact, provide all the data to the ideographic brains.)

One has to laugh at ideographic descriptions of what happens when the mythopoetic brains are finally able to break through the prison of the ideographic prison: serendipity, it's called! Why? Because we, as a species are rapidly losing our ability to call up these other brains at will. Imagine, for a moment, the power of our human capacities if only our ideographic brains would simply accept as legitimate, the many mythopoetic ways of knowing. Serendipity, indeed!

✿ ✿ ✿

Cultural Identity and the Mind/Mind Split

What can we truly call cultural identity today? Our national identities, ethnic identities, familial identities, socio-economic identities, matrimonial identities and even religious identities have become so obscured and compartmentalized that an individual is usually found floundering about from one identification to another, without any real understanding (let alone affinity) to any one of these constructs. Why? Because as the history of the heresy has shown us, most of our present day so-called identities are artificially constructed, external unities imposed from the outside on the many individuals. And identity today comes primarily from that individual's decision to accept the codified laws and beliefs of the CONCEPTUAL AUTHORITY upon which that artificial construct maintains its autonomy. In other words, we choose to believe the WORDS required of us to belong to that group and we act accordingly. Everywhere. Think about it. So where is our INTERNAL identity? Further, how can we learn to identify and cooperate with others' identities? Certainly not by the criteria of the mind/mind split: by collectively following the "impartial" law!

An historical survey of cultural identity reveals that what truly keeps a group together is its myths. Myth here is defined as "a large controlling image that gives philosophical meaning to the facts of ordinary life; that is , which has organizing value for experience."[14] In other words, cultural identity is biocultural identity; with each member of the group capable of reconstructing EXPERIENTIALLY the foundations of that group's unity, through the individual's exercise of his/her OWN neurobiological apparatus as the source of verification.

By unifying a society through the shared traits of our neurobiological inheritance, our internal identities then become authenticated and we, as a result, become again a human cultural community. This is not to suggest that we unify under a single mental property; that is what we have already done, with the results being our neurobiological slavery to the logos bioculture and its conceptions. Rather, what is needed is a global understanding of what we, as a species share AND how we, as a species uniquely diversify and communicate our shared inheritance to the benefit of all. How can we achieve this? Through education.[15]

Education systems that train children in expressing and exercising ALL the mental faculties inherent in our human repository is the only way in which true globalization can occur. (Again, globalization here refers to humanity's acceptance of our universal biological inheritance as well as the diverse methods by which we unfold them, NOT the unification of the all under the authority of the one!) Though individuals will certainly fall under a single biocultural affinity, by the acceptance of this diversity, no group or bioculture would gain superiority over the other, nor would we lose our collective biological inheritance for lack of exercise.

And finally, then, we can alter our human spiral from the throws of virtual reality as the greatest source of human sensation, to our own human bodies as the greatest source of our new Virtual Reality: Human Identity.

Conclusion: The Rise of History: The Fallacy of Phallicy

Introduction: The Clenched Fist of Ideography, or "Madam, I'm Adam"

We have certainly made it apparent in the previous sections of this chapter, that the effect of ideographic language, through its constant paradigmatic replication in other human substitution systems, brought about significant changes in human society. What we have yet to discover is the extent to which we, as modern members of this human society continue to be manipulated and controlled by this seemingly invisible web implanted within each one of us through this mindset. Has the mind/mind split become so much a part of what it means to be human that we, as a species have already become hardwired by its effects? Or rather, are we simply the unassuming victims of a very intentional strategy to destroy our human diversity in the name of some higher global unity, as imagined by an all-knowing few? Conspiracy theories notwithstanding, the true underpinnings of the mind/mind split lay in the balance of power amongst vying patriarchal factions over the rightful titles of legitimacy, passed on, as it were, through the authenticated lineages of the first fathers: the ancient Phoenicians, from whom men learned the power of the fist and of the phallus. How this happened historically is almost not as important as how this exact, universal hierarchical structure continues to hold sway in virtually ALL aspects of human public life, and even in a few of the aspects of what we (wrongly) perceive as our private lives.

The Universal Paradigm of Hierarchy or When the "Flies" that "Blind" are the Ties that Bind"

It has clearly been attested that the Biocultural Paradigm is one based virtually upon the structure and functions of our own human intelligence systems, evident in our neurological functions; therefore, it is safe to assume that this model is in fact the first example of a substitution system at work. In mythopoetic biocultures this understanding is not only known, but it is also the entire purpose of educating humans by that substitution system: to elucidate the communal faculties of the human mind, and assure the continuity of this transmission across time. It follows then, that the biocultural paradigm lacks an inherently universal hierarchical framework, since each of the acts represented by the respective biocultures merely represented one single aspect of the total array of human potentiality; one's individual biocultural affinity being based solely upon the choice of which function would serve as the primary organizer of that array of sensation. The introduction of ideography however, altered this understanding, either intentionally or unintentionally, due to the realities confronted by these ideographic communities of merchant traders, or of those whose homelands were no longer existent or no longer tenable due to climactic or other changes, (i.e., the Phoenicians); as they sought to establish themselves as stable "cultures" on foreign soil:

-the need to establish a method of tracing material wealth (including land and children) through paternal lines, so as to acquire dominion and control over conquered territories.

-the need to establish a method of maintaining autonomous control over the indigenous populations, whose members undoubtedly would outnumber these foreigners, while still maintaining the necessary balance of trade so as to insure basic subsistence (i.e., food).

-the need to reiterate the superiority of their own skills over those of the indigenous population, which in fact was not at all difficult, as these skills were indeed far superior in areas such as astronomy, building, metallurgy, etc.

-the need to universalize the language of obedience through the institution of the WRITTEN LAW.

The obvious method for achieving all of the above was found by the institution of what to us becomes the entire inversion of the biocultural paradigm, frozen as the universal paradigm of hierarchy. The word "hierarchy" itself means something like, "sacred origins" and that's exactly how the model was established by these logos founders. The biocultural model is our sacred human legacy, and has always been that; the model inverted and frozen, (inherent as it was in our own human psyches, albeit in fluid form), became the easiest method of establishing cultural hegemony through the establishment of the simulacrum of biocultural hegemony. In other words, the model was familiar to the populace; therefore, the subsequent ideographic perversion of it by the usurpers was almost child's play.

Let us review then, this new model of hierarchy, as established by the ideographers. In order of importance:

> LOGOS: The Chief God/Leader who prescribes laws and makes legal contracts with the indigenous populations.
>
> LEFT BRAIN MIMETIC: The Council of Gods/Lords who interpret and enforce the laws and ideals of the logos leader.
>
> RIGHT BRAIN MIMETIC: The priest/kings whose function it is to mediate between the Gods and the indigenous folk; to maintain the records of transactions, the secret knowledge of the leaders, the scientific discoveries of the new "culture" and basically keep the peace.
>
> MYTHOS: The former leaders of the indigenous folk who become re-royalized by marrying the daughters of the God/Lords thereby maintaining their "royal" status, though the indigenous folk usually end up hating them for "selling out."
>
> MAIA: The common folk who continue to live their lives, a little happier because some "big brother protector" will always be there to

> watch over them in times of need as long as they maintain obedience to him and his people.

And there you have it: the mind/mind split, frozen for eternity.

The Myth of Legitimacy, or What Came First, the Chicken or the Egg?

Once the hierarchy of ideography is established in a community, it is easy to maintain control of it because the lines of demarcation are in fact hardwired into our neurological inheritance, especially if our mythopoetic skills atrophy for lack of use, as is what occurs in such situations. How, though, you may ask, is this seemingly absurd hierarchy established in the first place, if the people are originally mythopoetic? The answer? The myth of Legitimacy, which avers that men are the source of creation, through the medium of their "seed" and that women are merely the "entombed" receptacle in which the seed becomes reborn. With this myth, ideographic tyranny exploded. For after all, if a masculine force can prove palingenesis to a mythopoetic person, believe me that mythopoetic person would surely deem them superior. In fact, the ones who would first bow down prostrate before that force would be the masculine members (no pun intended here) of the mythopoetic societies, for obvious reasons. And if one were to add to this the promise of ownership of land, wealth and heirs to worship them, making them immortal, mythopoetic men would sacrifice life and limb for the opportunity; in fact, they did.

When one looks to the mythologies of these ideographic usurpers, one can immediately begin to see the strength of the effect of logical substitution systems on humans. Oddly enough, the totemic icons of these groups reflect

EXACTLY the intentionality of maintaining masculine rights of inheritance, due to the masculine, not feminine roles in the creation of humans.

Consider the following symbols: Birds (especially the eagle and phoenix); the winged solar disk; the serpent; the sacred trees; the butterfly; the fish; the sea turtle; the sun; gold; the crown of royalty; the crimson robe; the sceptre. Of the symbols referring to animals, do you see a common thread? The answer is that all of these animals are CREATED without a mother. Butterflies, a very popular soul symbol, are hatched from a chrysalis, and prior to that were phallic little caterpillars; serpents are sacred for the same phallic reasons. If you know anything about mythology, the history of this transformation will very quickly begin to make you very uneasy, especially when you analyze the next group of symbols, namely, birds, fish and turtles, all of which are hatched from eggs. Consider for example, the cosmic egg; the eggs from which the children of Zeus and Leto were hatched; Protogonos, the Orphic demiurge who was hatched from an egg; not to mention Easter eggs. In addition, that birds are the sacred icons of kingship, remember the infamous "Birds" of Aristophanes, where he parodies the mindset of Phoenician merchants, and their "gods." Recall how Ovid, in his "Metamorphoses" thoroughly satirizes all the bird myths formerly sacred to these ideographs (and this was probably the true reason for his banishment from Rome). And finally, now we can truly comprehend the full import of "flipping the bird" and why the "cock" is sacred to the sun. Trees are phallic progenitors as their sprouts are created without female interaction at all. Likewise, are plants and grain born without the aid of female assistance as the seed of the sun god descends from the sky as rain to resurrect the "dead" seeds which were previously entombed deep within the barren earth. For, without the seed of the father and/or the seed of the dead masculine phallic looking vegetation (corn, wheat, etc.), no earth would produce worthy harvest for humans, right? Finally, the sun, as the ultimate father without mother, is the model par excellence of ideographic "culture." In fact, the gold crown of kings IS the holy symbol of the sun, gold being the most precious metal because of its resemblance to the high god. The crimson robe of kings represent the life-giving blood (OK, so it's really red from the dye of murex snails, but to the Phoenicians who invented it, red is, after all red.), and the sceptre is the rod that never "dies."

In short, what these ideographic usurpers did was to "instruct" the indigenous people in the real knowledge of things, by introducing those elements into myth that would in fact strengthen, in the eyes of the MYTHOPOETIC folk, the veracity of the claims of the ideographic usurpers! In short, the left brains pulled a fast one, as only they could do, and it worked.

Hurling the Final Blow: From the God of Abraham to the New World Order

In the third chapter we outlined what we termed the "logos hero" paradigm and traced it, mythologically, to its roots in the creation hymns of the Rg Veda. Recalling these two paradigms now, we shall examine how once again they appear, not however as myth, but as a totally refurbished system due to the introduction of the hierarchical context inherent in the adoption of the mind/mind split. This new system is what we have come to know as history: as the chronological transmission of events concerning and tracing the lines of legitimacy amongst peoples, nations and ideas.

To reconstruct this origin of history, let us first recall the mythopoetic context of the logos hero model, to comprehend exactly how and from where each of the subsequent moves of history originate. We shall follow with an interpretation of the ideographic shift and finally, offer examples *from history* which support the contention that everything we as modern readers may view as claims of legitimacy; be they human or divine, are merely ideographic substitution systems of one basic mythopoetic piece of information, that being, the structure of the human minds.

In the Rg Vedic creation scenario, two modes of knowledge were outlined; the knowledge of the "seers" and the knowledge of "man." These two modes of knowing were both created in the ASAT, that is, in the realm of potentiality, before forms were conceived.

In this model, the path of the seers was initiated by "rushing waters" and "the light of the sun." The path of men was initiated by the "bullcow sphere" and by the god Indra breaking it open. (See appendix for a review of the original myth.) These two paths, viewed from the mythopoetic perspective reveal two ways of "seeing." The second way of seeing is the one most familiar to the modern reader; it is the seeing of objects as separate entities of form in distinct space/time. The former way of seeing poses more problems, for *it* is the way of seeing by continuity before manifestation; therefore, it hinges on the act of form making rather than the cataloguing of forms made. The mythopoetic mindset allows for the experience of both of these perspectives; the ideographic mindset allows for only the second way of seeing, that is, by differentiating objects made, and herein lies the crux of the issue regarding the transmission of *this* model of creation across biocultural boundaries. If the seers see by the path of the gods because the gods flow through them (meaning: the seers are able to create worlds by the acts of imagining. The gods here represent archetypes of *actions* which, repeated, manifest themselves in the material world; hence, are "born"). then the "seers" in some ways become as gods (meaning: know the path from immortality/potentiality to mortality/manifestation). Read ideographically, however, this *concept* of being as a god, takes on entirely new meaning. For one thing, ideographic biocultures function through the substitution system of the self which is the epistemological invariant for their world to exist, so immortality for them is the immortality *of that self.* The incessant search for this immortality is an obsession with ideographic biocultures for the obvious reason that if one focuses on form, including one's own, then it is blatantly evident that eventually *it* will cease to be. The search then for this immortality leads ideographic mindsets to posit certain criteria, albeit external, to assure the immortality of that invariant self. And here is where and how the myth of legitimacy gains veracity, at times with deadly consequence for many.

The logos hero paradigm is the clearest example of the turning point from myth to history, for it is a unique combination of an attempt to catalog historical figures through the mythic criteria of the Rg Vedic creation myth (re-told, of course) in order to make these great men "as gods," in other words, immortal. The problem of course is that they have no idea what the original intention of the myth was, though the *reading* of the story is more or less comprehensible ideographically. The ideographic reading is something like: a male who is somehow created out of the sea and is the child of the sun, who travels through the wilderness and encounters weird forms like monsters, and who becomes king but eventually dies on some mountain top and is whisked up to the sky to become a god after he dies. Unfortunately though, the mythopoetic reading is this: before birth and after death represents the world of continuity, as exemplified by the amorphous qualities of water and fire ("running waters" and "the light of the sun"). Discontinuity occurs when one is born into material creation and is given a form through the breaking apart of the "bullcow" into "bull" and "cow"; in other words, into separately identifiable objective realities, including selves. The path of mortality is grounded on the perspective of men; that of focusing on "mere form." Conversely, the path of immortality is achieved by focusing on "the power of creating forms" through the acts of mythopoetic creation. *Thus immortality can be achieved while still alive; in fact, one needs to be alive to achieve conscious awareness of this act in order to ever become "as a god" in this model.* Of course, upon the dissolution of this mere form (the death of the biological *or* biographical self) the cosmic power of creation will return to the state of continuity from where it originated until manifestation will again occur; in which case, the perspective of discontinuity and external reflection will again be apparent.

Imagine now what the ideographic usurpers did with this story. First, by focusing only on form and with an end towards the immortality of the biographical self which *is* the left brain without the body, (the only item worth remaining immortal), the substitution system of creating a model for achieving immortality was of primary importance. The easiest way of achieving it was to become "as a god" of course, which in ideographic terms meant that the self would live on beyond the lifespan of the bodily capsule. So they turned to the myth of creation which outlined the path of the gods and they found the story of how immortality was created by the forces of the sunlight and the rushing waters; so all heroes were children of the sun and were thrown into the waters of the sea at birth. Since, in the Rg Vedic creation myth, the "unmanifest one foot" was the first thing to move in the waters of the ASAT, which precipitated the creation, it is no small wonder that that "one foot" became the phallic symbol that was responsible for making all things. From there, it was not difficult to re-create the entire myth to explain ideographically how creation and immortality came out of an original phallic power, and how each male had the potential for immortality through the use of that phallic power, especially if he were descended from the lineage of the "immortal" path; that is, of the gods. And since both paths were descended from a common origin, the uncreated one foot, then it was only the *lineage* that would determine whether one were a child of immortality or a mere form. And herein lies the first inkling of the hierarchical structure taking form.

The closest to immortality were the children of the sun (kings) and those who imitated the bullcow by retaining the powers of the opposites within themselves (priests). These formed the hierarchical lines of descent, with both views having their own methods of achieving immortality (to be discussed later). Under the kings were the divine lineage sprung out of the king's union with a woman of similar background. Under the priests were the oldest sons of the kings whose job it was to eventually leave the kingly household to set up his own. (This hierarchical line is equivalent to that of Yamayami and Yama.) The third level of hierarchy are the mercantile and agricultural folk; those under the king's lineage being the former and under the priestly lineage being the latter. These classes correspond to the role of the Asvins in the Rg Vedic myth cycle. And finally, those for whom little hope of immortality exists, save the outside opportunity of having been chosen to become immortalized by sacrifice, remain forever on the cycle of mere form as only that, mere form. They are the lot out of which the priests draw candidates for "salvation" (read here, they are the sacrificial victims; sometimes, literally.) In India, the introduction of ideography led to the caste system, with all the social baggage included. But a much more serious and devastating thing occurred elsewhere with the introduction of this model, via the ideographic biocultures; the widespread institution of human sacrifice, the epitome of which *to this day* is heralded as *the model of human redemption and immortality*.

The mythic account of the logos hero as recounted in the third chapter outlines the basic story of those leaders in myth who were known to be founders or leaders of great houses of royalty. In other words, they became, in the myth cycles, heroes who ascended to the realm of the gods. For example, we recall how Oedipus was mysteriously called up to the heavens, how Perseus became immortalized in the stars, and how Theseus became worshipped as a demigod after the Athenian victory at Marathon. In each of the above cases, the myth cycle ended mythically, for it is assumed that these characters achieve divinity mythologically as their existences are likewise assumed to be archetypal and therefore mythic, as well. But a re-reading of this logos hero paradigm in light of its appearance elsewhere in the ancient and modern world, yields an interesting argument in favor of interpreting these stories not with mythopoetic understanding, but as a quasi-mythopoetic paradigm assigned posthumously to actual, historical figures. In other words, this logos paradigm, it seems, is the posterior mythos used by ideographic peoples to relate the fact that certain men were to be deemed gods (in the Phoenician sense). This interpretation is evident in scores of examples from history. Beginning with the earliest examples of characters whose lives follow this paradigm, we are confronted with the Egyptian Osiris, whom the Egyptians clearly depicted as having existed in historical times. Following his human model, the Egyptian *Pyramid Texts* and later, the *Book of the Dead*, recount the ritual of the deceased pharaohs to become immortal by following once again, the path of the logos hero, albeit, in logos form. We have already noted that Moses' biography follows the paradigm of the logos hero; what we must now add is that the entire account of the *Exodus* re-creates for the entire chosen population of Moses, the same path as that of the logos hero. In other words, the chosen people are destined to become immortal by the fact that the story of the flight from Egypt is in fact a re-telling of the logos hero paradigm for the communal group. Sargon ll of Akkadia was destined for immortality by the telling of this paradigm in his biography, as was Romulus in Rome, Ion in Ionia, Karna in Bengal, Erichthonius in Athens, Perseus in Mycenae and

Tiryns, Oedipus in Thebes and Joshua in Canaan. These founders then formed the lineage of ROYAL LEGITIMACY in those geographical areas, and therein began the myth of legitimacy by divine decree.

Turning our attention now to more recent history, let us consider two early characters from the common era: Judas[16] Iscariot and Jesus of Nazareth. The biography of Judas clearly follows the life of the logos hero: he was sent out into the sea in a chest, which was washed ashore in Scariot. The queen there raised him. Then he went to work at the courts of Pilate. He killed his father and married his mother. And he had a scapegoat to serve as the sacrificial offering in order to assure that the new kingdom would in fact be established. That victim, was Jesus. And now we are faced with the real question: who then IS this Jesus and where is his place in this history? Was his divine appellation due to his surrogate position with regard to Judas' use of him in this ritual? In other words, has the story of Jesus historically come down to us with this mythos attached precisely to aver, to those followers of the mind/mind split, the authenticity of the messianic prophecy of a NEW KINGDOM ON EARTH? Or perhaps the historical Jesus was the next in the line of royal succession to become the *de facto* ruler of his peoples, as his royal lineage might reveal. Was he then murdered by those in charge (the Romans? Herod?) because of THEIR fear that the lines of legitimacy might shift to another ruling house? Or perhaps the historical Jesus was a masculine wonder-worker whose pro-masculine powers of creation were heralded as a model for all men to follow. Perhaps his history is to reveal the importance of masculine supremacy through the esoteric teachings of "masculine knowledge"; teachings that he gave ONLY to the elect twelve (and St. Paul, via telepathy!); to be transmitted only to those men in the proper line of succession from him or his followers. (Read here, the lineage of the priesthood.)

In any of the above scenarios, the message still rings clear: he was a mortal who ascended into the divine realm after his death and became god. And his function was to establish the foundations of a new kingdom on earth. Isn't that the message, short and sweet?

Continuing on in history, then, let us turn our attention to the one man who, more than any other is considered responsible for bringing this new kingdom, Christendom, to its heights: Constantine the Great.[17] The truth of the matter is that it was not Christianity that Constantine instituted, but rather the Syrian (read here, Phoenician) cult of the sun god as the One god, as the appellation, "sol invictus," which the buzz word of his reign, indicates. His tolerance of Christianity, as well as other similar monotheistic beliefs, (such as Mithraism) rested upon the SIMILARITY amongst these cults. (Consider, for example, how because of Constantine, Christianity heralded Sunday as the holy day as opposed to Saturday, which they had previously observed as the Sabbath). But when it came to choosing a royal line of legitimacy, Constantine did choose the God of Christianity; however, HE USURPED THE ROLE OF JESUS, THE MESSIAH, FOR HIMSELF. He, thus, inaugurated the possibility, once again, (or still) of a human's capacity to become deified by establishing a kingdom on earth; in this case, the empire of Rome. Constantine took Jesus as being merely a human manifestation, as he himself was, of Sol Invictus.

So, what of the mythos of Constantine's life? We are told that he received his call to greatness in a vision of the sign of the cross appearing in the sky with the words, *in hoc signo vinces*, written upon it. This sign, legend says, led him to defeat his enemy, Maxentius, who was also in line for the imperial throne. After his ascension to the throne, he killed his second wife (the first one disappeared) and first born son by his first wife, because of his alleged fear that should his son become involved with his step-mother, Constantine would lose the right to the throne. Curiously though, his son became venerated as a blessed martyr (a sacrificial victim, perhaps?). But the Church fared well under Constantine, due to the emperor's attempt at unification among all monotheistic churches and the state. But after his reign, the fight for legitimacy once again surfaced. The only thing that changed was that this time, the Church itself was to play an even more significant role than it had during Constantine's reign: the role of DIVINE MEDIATOR among warring factions for the throne. In other words, the descent of legitimacy from the lineage of "men turned gods" had become transformed to the descent of lineage by God's decree- as expressed by His spokesmen on earth: the Church itself. It is no small wonder then, that the major houses of Royalty in Europe still trace their descent and lineage, granted by God through decree of the Church, with the oldest Royal Lines being descended from the Constantinian branch itself.

Thus this myth of legitimacy continued through to the time of the Protestant Revolution, when a new nobility, tired of the manipulations of the Church, sought to re-introduce the lines of succession by the criteria of another, albeit as ancient, line of legitimacy. They found that line in none other than the ancient Phoenicians themselves, and the organizations that promulgated its credibility became the new mediators of god: the Secret Societies.

Through the re-introduction of the ancient rites of "masculine knowledge," now founded on the ancient principles of alchemy, magic and gematria, as promoted by the emergence of the new wealthy mercantile class of Europe, Phoenician ideals were once again peddled as the criteria for the "elect." It is odd that at the root, all of these seemingly competing orders, religions , political factions and societies were in fact founded on the same mind/mind split; unfortunately, that factor is useless when the object of the group is to gain global hegemony precisely against those other groups, in the same way that their "historical" predecessors had . So we recount in history such events as the Age of Reason, the Age of Enlightenment, the Age of Exploration, the Age of Revolution, as isolated phenomena in a seemingly chaotic stream of events, rather than as systematic moves created by the few as dictated by a plan that was already mapped out. We see the discovery of the New World as a stroke of luck by some ingenious sailor who had no idea of even the dimensions of the earth, rather that seeing what it really was: an attempt to REGAIN what had been lost in history. And finally, we are led to believe in the myth of progress, that the future holds the dreams of our lives fulfilled, that upon our deaths we will be resurrected and become immortal, that the more we suffer at the hands of oppression, the greater our rewards shall be in the future; in short, we've bought into the dream of the mercantile

Phoenicians. Unfortunately though, since the acceptance of the ideal of progress is merely the historical evidence of the mind/mind split, and nothing else, and since all aspects of our culture have been infiltrated ONE WAY OR THE OTHER by this mindset, it is difficult, first, to SEE it, but more importantly, to fight it.

Perhaps that is why the millennial murmur of the end of the world is upon us again: because we can't seem to get our minds out of that mind/mind split and logos loop. What should be done? Our ideographic answers are clear:

-build an ark and hope for a flood (Fundamentalist answer)

-discredit the messenger, once and for all (Humanist answer)

-kill your father and marry your mother (Feminist answer, just kidding!)

And the mythos response? Ready yourself for the return of the Golden Age, for the Iron Age is surely upon us!

CONCLUSION

It has been a long journey from that valley in Naucrates to the vestibule of twenty-first century America. Many a nation has risen and fallen; many a war has been fought; many gods forgotten and others born; many languages, literatures, artistic developments, architectural wonders, technological advancements, agricultural achievements and scientific discoveries have come to pass. And still, we seek, as they did in ancient Egypt, the answer to those timeless questions: When will we, as a species become truly wise? When and how can we achieve the ultimate gnosis? And with these questions still left unanswered, the species moves forward. It is, in fact, our most basic impetus, isn't it, to know?

What we've come to realize with the understanding of the biocultural paradigm is that knowledge; that is, what constitutes knowledge, is bioculturally determined. In other words, each bioculture creates its own primacy of one aspect of human possibility as its ground for knowledge, as defined by that bioculture. And collectively, all the knowledge from all the biocultures, for us, as a species, should be what constitutes true wisdom.

The knowledge of the maia bioculture, with its origin in the reptilian brain, is the knowledge of sexual reproduction. Thus, the primacy of the role of the sexual encounter forms the unifying image of that bioculture: the myths of creation re- establish the movement of sexual energy towards procreation, the eschatological system emphasizes the constant movement to and from the sexual origin (mother, the One) and the arts, the social organization, and overall human pursuits develop upon the preservation and control of this knowledge: the knowledge of reproduction.

The knowledge of the mythos bioculture, with its origin in the limbic system, is the knowledge of homeostasis and harmonia. It is the knowledge of how divergent impulses can work together and establish a biological unity greater than the vying forces in opposition can. Thus, the unifying images of the mythos bioculture are the archetypal forces, the "imageless" forms which govern the parameters of our material lives: the gods. By each god representing a single human act necessary to maintain the homeostasis of the collective group, the bioculture of mythos assures its continuity. And the knowledge, then of this bioculture is the knowledge of how so many seemingly divergent forces can be accommodated to form a unique whole, through balancing the opposites to create a new synthesis among humans; the same way that our limbic system creates this unity within each human.

Right brained mimetic knowledge is the knowledge of form-making; of reformulating and translating the existing mythos impulses in such a way as to create new realities and new possibilities. This is the bioculture of the demiurge, who is able to replicate, by will and memory, the geometric structures of the mythos bioculture to make separate images of these structures in order to observe them. It is this bioculture that gives us the sense of subject and object; heaven and hell; here and there; not as opposite forces to be reconciled but as choices for material realization in themselves. Here, for the first time, a bioculture can create a hierarchy of objects with itself at the top, based on the primacy of the sense of perception, because the observer of the visible phenomena, in the act of observing becomes the temporary invariant form in order to make possible this vision. Hence, the demiurge image.

So with right brained mimesis, the construct of the human self is born. And with it, society moves to another level altogether.

In communities where the demiurge god is viewed as a sort of magician/craftsman, the bioculture remains right brain mimetic. When, however, the demiurge god is the celestial tyrant, and humans become the mere slaves to his almighty will, the bioculture has moved to left brain mimesis.

The knowledge that the left brain mimetic bioculture gives us is the knowledge of yes/no; the knowledge of choice and as a result, the knowledge of internal conflict; for its biocultural primacy lies in its mimesis of the left hemisphere of the neo-cortex. And as we have learned, the left hemisphere takes upon itself the role of image police-a role not entirely negative, as it is this function which allows us humans to recall, from stored images (as concepts) in the left hemisphere, past experiences that bear similarity to present ones; thus, enabling us to make decisions in the present based upon experiences in the past, through the medium of concepts. In and of itself, this knowledge is extraordinarily useful. Likewise, due to the knowledge of this bioculture, we can also know symbolically, and this, too, adds unparalled strength to the human species. BUT, what happens when the concept maker usurps the power of the image makers? What happens is the mind/body split. The body (and its ways of knowing through image-making) becomes evil and the left brain, conversely (and its way of knowing through concepts) becomes the dictator. And the battleground becomes the human flesh, because the body thinks the voice of the dictator is an almighty god OUT THERE SOMEPLACE.

The uniquely genial logos bioculture, also originating in the left hemisphere, was created to alleviate this mind/body tension by diverting the conceptual systems outside of the body and forming substitution systems based on concepts (which do not need the medium of the other brains, consciously, at least) as the origin of its system of knowledge. Here then, the functions of the left hemisphere become thc model of knowledge and this bioculture then functions by replicating and improving IT in the material world.

So this bioculture then uses the left hemisphere activities of concept making as the source of knowledge, but for the purpose of discovering the universal formula for objective knowledge; that is, KNOWLEDGE UNCONTAMINATED BY HUMAN ERROR. (Read here, knowledge that could not possibly contain any human subjectivity, or it would destroy the veracity of the find.)

And so, our search for knowledge comes to a close. We, as humans want to KNOW without being, ourselves, any part of that knowledge. And there you have it: the mind/mind split. Alas.

But what of wisdom, in all this knowledge? I think my father said it best, when he answered this question, posed to him, as it were, by me, the child who just had refused to go to school one day and was fighting for a reason ever to return there.

"Maria," he said, "wisdom is what forces us to carry on the search, long after all the knowledge is removed from us by life's experiences."

Thanks, Dad . . .

APPENDIX 1.1

THE TRIUNE BRAIN

(after Paul MacLean)

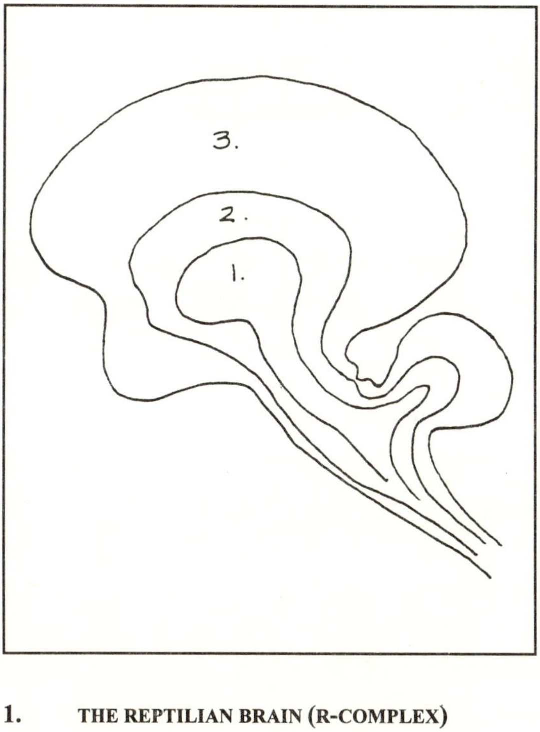

1. THE REPTILIAN BRAIN (R-COMPLEX)
2. THE MAMMALIAN BRAIN (LIMBIC SYSTEM)
3. THE NEO-MAMMALIAN BRAIN (CEREBRAL CORTEX)

APPENDIX 1.2

THE BODIES HUMAN AND THE BIOCULTURAL PARADIGM

Introduction

The Biocultural Paradigm shows that the human body is conditioned and thereby functions, as an image of the body imposed upon it by the bioculture operating upon it. As the examples on the following pages display, each bioculture views the human body image by the criteria of the primary sense of that bioculture. Thus, the Maia Bioculture diagrams the circles of energy moving from the reptilian seat at the base of the spinal cord, up through these power centers to the crown of the head. The images are: the serpent (kundalini) and the energy circles (chakras), moving the mass of matter and transforming it.

The mythos bioculture divides the body as one would musically/mathematically divide a string; the genital area marks the OCTAVE (the body's replication center; the mathematical center of the body/string); the navel (liver area) marks the phi (Golden Section) going down, and the heart area marks the phi (Golden Section) going up. These are the mathematical proportions, but they are also the musical proportions of the FIFTH and FOURTH, respectively. These are also, remember, two of the three divisions of the tripartate soul of Plato and the Pythagoreans (the liver being the appetative seat and the heart being the emotional seat; both of which comprise the sources of thymos in an individual. Of course, the HARMONIA of these two sources of thymos determines the fate of the soul.)

In Right brain Mimesis biocultures, the model of the universe is superimposed on the body of man, for man is composed of all the elements in the universe; therefore, all the elements have "complementary" effect on man. This reasoning originated by mimetic bioculture's recognition of the effect of the moon on womens' bodies, and then theoretically extending this belief, to include the effects of ALL the planets, stars, constellations and intergalactic forces on men's bodies. For after all, if women had the moon, then men had the sun, planets, stars and every other celestial element "pulling" for them! Thus, the human body became a microcosm of the physical universe, which is the macrocosm.

In Left brain mimetic biocultures, the opposite is true. Here, the model of the HUMAN BODY becomes the ORIGIN of the structure of the created material world, for if the human body is a microcosm of the universe, then the human body should likewise be a Macrocosm of its own created forms in the material world. Thus, architectural floorplans for homes of the gods became modeled, not upon the geometric proportions of the archetypal world, but upon the HUMAN BODY PROPORTIONS. Man, that is physical man, becomes the measure of all things in the material world.

In the Logos bioculture, the physical body is merely a cumbersome appendage to the real human, which is the imageless brain. Thus, its parts, like the parts of a car, can be altered, cut out, replaced, by new ones, and never affect the TRUE PERSON, because the true person has nothing to do with his/her body.

Chart Displaying Major Chakra Centers in Body.

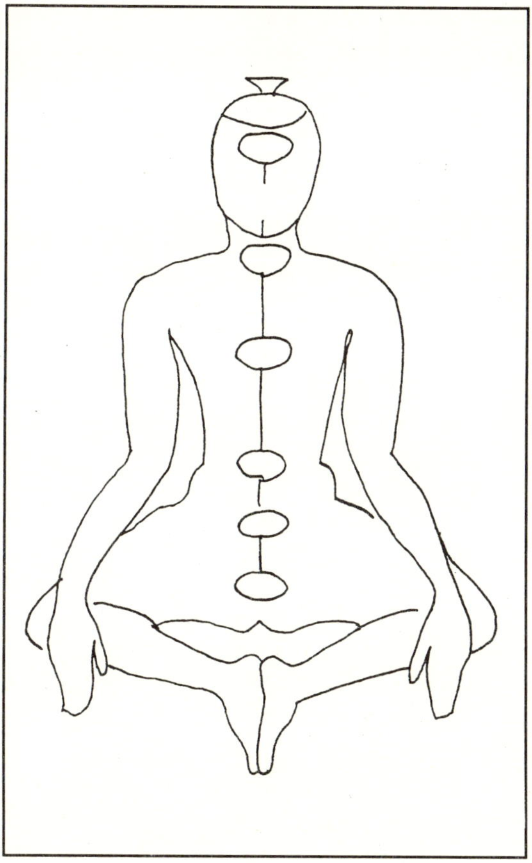

The Mythos Body

Chart outlining the body as a musical string with the following centers segmented: head/feet = fundamental tone; reproductive area = octave; liver area = musical fourth/reciprocal fifth; and heart area = musical fifth/reciprocal fourth.

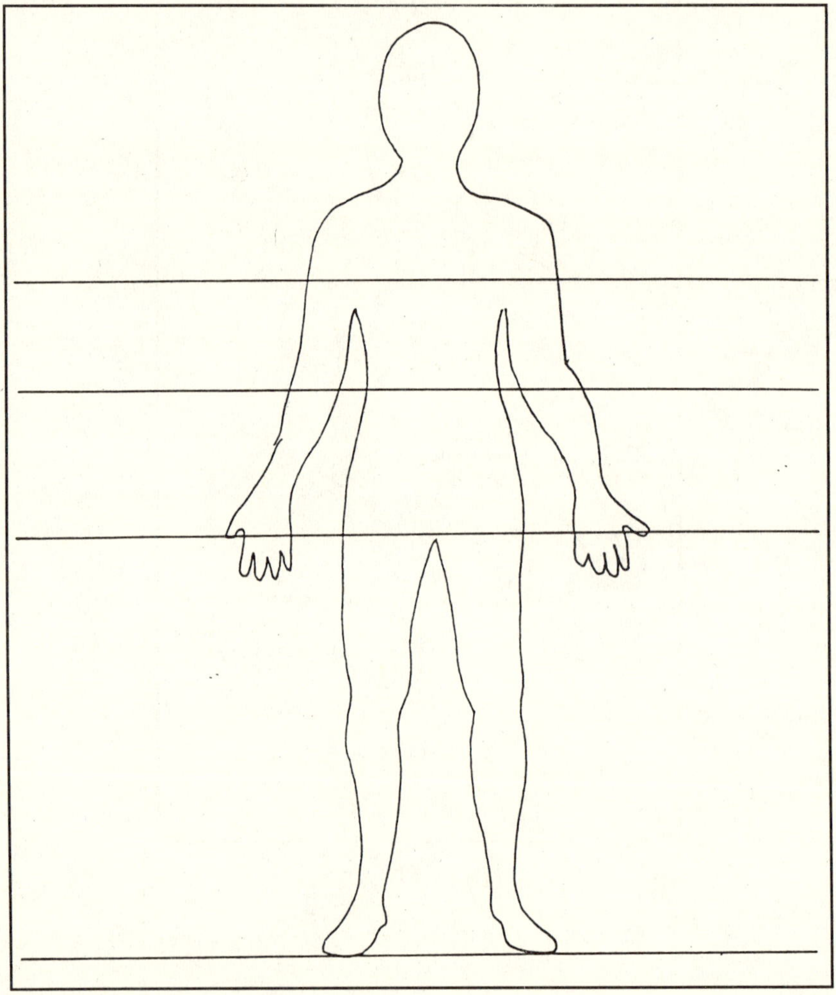

The Right Brain Mimetic Body

Chart placing the location of the zodiac signs at certain points on the human body (after 14th century German astrological mss.): Aries, head and face; Taurus, neck and throat; Gemini, arms and shoulders; Cancer, breast, stomach, and lungs; Leo, heart and back; Virgo, belly and guts; Libra, hips; Scorpio, bladder and sexual organs; Sagittarius, thighs; Capricorn, knees; Aquarius, legs; and Pisces, feet.

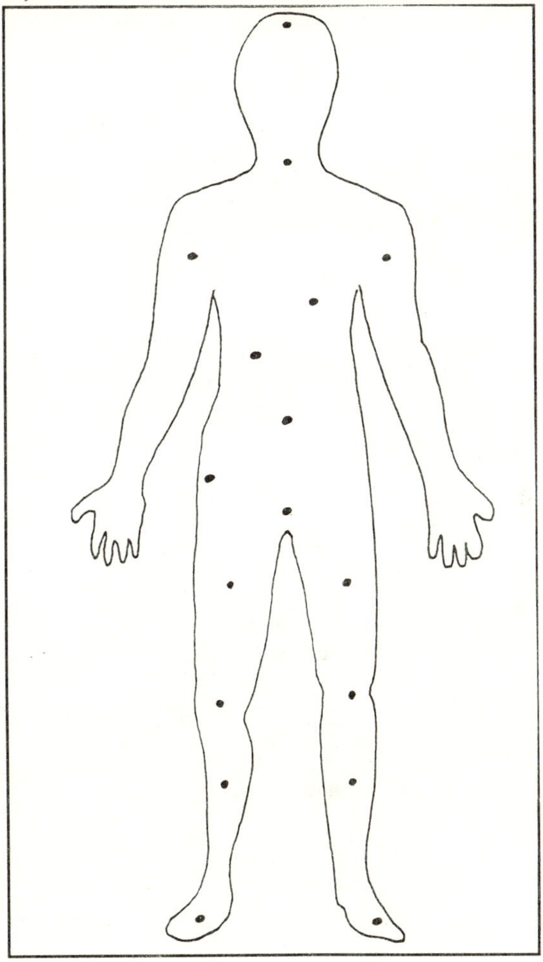

The Left Brain Mimetic Body

These architectural "erections" bring new meaning to the ancient Greek aphorism, "Man is the measure of all things."

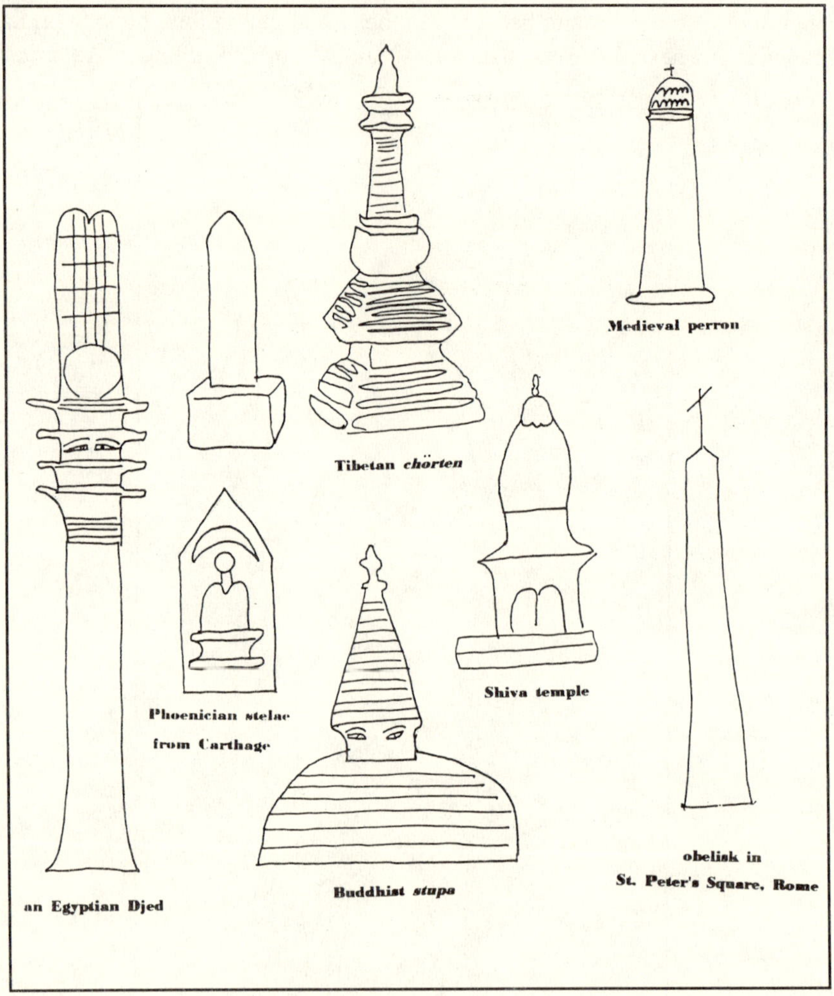

Chart Delineating the Major Organs of the Human Body

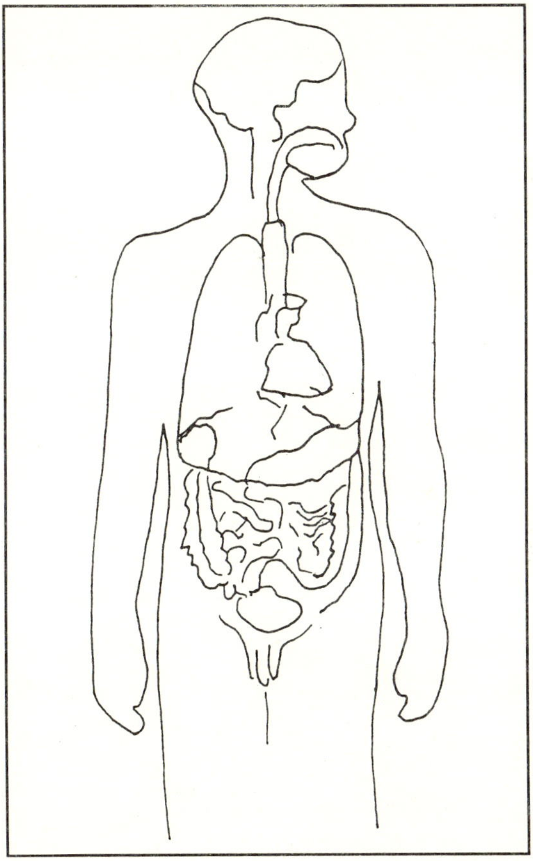

APPENDIX 2.1

Eschatology and the Biocultures

Introduction:

In the same manner that each bioculture creates a specific cosmological order, and one is able to discern the bioculture from the cosmology; each bioculture develops views of the afterlife of the individual. These eschatological perspectives appear to shed the most light upon the veracity of the biocultural paradigm discussion, because the subject lends itself, per se, to a strictly imaginative mental construction, which curiously enough, mirrors the cerebral movements of the respective biocultures. That is to say, the views of the afterlife of each respective bioculture marks an absolute affinity with the workings of the dominant brain portion of that bioculture.

What follows then is a brief introduction to each of the views of the afterlife as developed by each bioculture, followed by a representative textual memory from that bioculture to support the interpretation.

The Maia Bioculture and the Afterlife

The infinite struggle between motion/nonmotion, presence/absence evident in the Maia bioculture is described in the cyclical model of the afterlife, where all forms are born of the goddess and eventually, return to her in order that re-creation take place. The necessary data useful for understanding this model is to focus on the action/inaction phase rather than the change of manifested form, as its significant message. Form is to be transformed because it breeds stagnation and stagnation follows the path of ASAT (inaction) not RTA (eternal movement); therefore it is wrong. Thus, in a Maia bioculture, it is irrelevant whether one's biographical self be "saved" or soul be made immortal; in fact, that would be disastrous, as all must retain fluidity for transformation to occur. In fact, in Maia biocultures, there is no body/soul distinction because there is no invariant eternal human.

❁ ❁ ❁

Textual Memory of Maia Bioculture

You are the original of all manifestations; you are the birthplace of even us; you know the whole world, yet none know you . . . you are both subtle and gross, manifested and veiled, formless, yet with form . . . Resuming after dissolution your own form, darkformless, you alone remain as One ineffable and inconceivable . . . though yourself creator, protector and destroyer (Mahanirvanatantra).

The Mythos Bioculture and the Afterlife

Perhaps the most interesting and also misunderstood view of the afterlife is the mythos view: Homer's morbid image of emotionless shades wandering about for eternity in an abysmal plain hopelessly awaiting some sacrificer to remember him so that he might once again feel, leaves the modern reader with an almost existential despair for the poor Greek whose afterlife this was. Fortunately, though, by properly comprehending this view of this afterlife in light of the biocultural paradigm of mythos, we can begin to not only cease our piety, but actually feel remorse for those who are not destined to become "among those dreaded shades."

How does a bioculture create such an eschatological view as this? Basically the model is as follows: Each human possess a matrix of form (geometric) which is then, during life, endowed with motion/emotion (thymos), breath (psyche) and divine intellect (noos). Physical form, then serves as the temporary harmonia of these vibrations in its matrix, much like a musical instrument, when being played, possesses harmonia in the music through the instrumentation of the form of the instrument's body (The nature of vibrational sounds is contingent upon the geometric proportion of the instrument itself.). Upon dissolution of the individual physical instrument (body) the harmonies dissipate, though the geometric form of the instrument is always retained as an archetype to be recreated, at which time, the harmonia will again return in the same proportion, assuming the geometric form is consistent.

A shade, then, is that form matrix of an individual who has once lived but has now dissolved its harmonics. By another living person lending his/her thymos to the matrix, that matrix is able to once again reverberate by the other's harmonia, thereby being able to "live" again, AS LONG AS HE/SHE IS REMEMBERED. Thus, the need for prayers for the dead. The emotion that causes harmonia of the thymos of a live person to "vibrate" in a dead person's matrix is EROS (love).

It is interesting to note that it is here in Mythos biocultures that a distinction begins to be made between body/vibration-matrix/movement. It is also here that the greatest issues of the nature of the soul are debated. And curiously enough, there are always at least three souls discussed. Could these "souls" be the vibrational patternings of the kinesthetic, auditory, and visual brains? A look at the following quotes prove that this is a distinct possibility:

> (Pythagoras) conceives the soul as comprising three parts – reason, intelligence and passion. The latter two are possessed by other animals as well, but reason men is in men alone. The bodily seat of the soul extends from the heart to the brain . . . the heart (houses) passion, while the portions located in the brain are reason and intelligence . . . Reason is immortal, all else is mortal . . . The rational powers of the soul are of the nature of breath; they and the soul itself are invisible in the same way air is invisible . . . When cast out upon the earth (after death) the soul wanders in the air, like body . . . Hermes . . . guides the soul by land and by sea to the uppermost region whereas the impure are not allowed to come near to the pure nor even near one another, but are bound by the Furies in unbreakable chains . . . (Diogenes Laertius)

> . . . there are three kinds of soul located within us, having each of them motions, . . . (o)ne part, if remaining inactive and ceasing from its natural condition, must necessarily become very weak, but which is trained and exercised, very strong. Wherefore, we should take care that the movements of the different parts of the soul should be in due proportion . . . (Plato, Timeaus)

Textual Memory of the Mythos Bioculture

> Our ship came to the farthest realm of deep-flowing Oceanus, where the country of the Cimmerians lies shrouded in cloud and mist. Bright Helius never looks down on them with his rays

> either when he ascends to starry heaven or returns to earth but dire night covers these poor mortals. Here we beached our ship and after putting the animals ashore we went along the stream of Oceanus until we came to the place that Circe had indicated. Here two of my men, Perimedes and Eurylochus, held the sacrificial victims, and I drew my sharp sword from my side and dug a pit about eighteen inches square. Around it I poured a libation to all the dead, first with a mixture of honey and milk, then with sweet wine, and a third time with water; over this I sprinkled white barley. I then supplicated the many strengthless spirits of the dead, promising that once I had come to Ithica I should sacrifice in my own halls, a barren heifer, the very best I had, and heep the sacrificial pyre with the finest things and offer separately to Tiresias alone a jet-black sheep that was outstanding among my flocks. When I had finished entreeting the host of the dead with prayers and supplications, I seized the victims, cut their throats and their dark blood flowed into the pit. Then the souls of the dead who had departed swarmed up from Erebus: young brides, unmarried boys, old men having suffered much, tender maidens whose hearts were new to sorrow, and many men wounded by bronze-tipped spears and wearing armor stained with blood. From one side and another they gathered about the pit in a multitude with frightening cries. Pale fear took hold of me and then I urgently ordered my companions to flay the animals which lay slaughtered by the pitiless bronze and burn them and pray to the gods, to mighty Hades and dread Persephone. But I myself drew my sword from my side and took my post and did not allow the strengthless spirits of the dead to come near the blood before I had questioned Tiresias. (Homer *The Odyssey*)

The Right Brain Mimesis Bioculture and the Afterlife

With the bioculture of Right Brain Mimesis, the emphasis of the afterlife concerns form; specifically, what form an individual will become in the next incarnation. In Right Brain Mimesis, as in Mythos, there is still emphasis on the separation between form and vibrations, but there are two major distinctions between right brain mimesis and mythos understandings of these concepts. First, while both agree that in life an individual is capable of "modulating" his/her vibrations in such a way as to alter his matrix, there is no reincarnation for a mythos person, but right brain mimetic people modulate specifically for their next incarnation. In other words, while in this form, one is training his or her individual vibrations to, upon this dissolution, reincarnate into another, hopefully higher form. This hierarchical distinction leads to the second main difference between the mythos and right brain mimetic biocultures on this issue; namely, in right brain mimesis, there is a birth of an invariant "soul-self" which takes on other forms, unlike a maia bioculture's reincarnation where no invariant form at all survives, for the most part. Thus, identity now moves from the matrix form to the vibrations (soul). And this shift now opens the door to the next eschatological view which will combine these views and assert the eternal life of both matrix and vibration, in the idea of the resurrection of the flesh and soul.

❁ ❁ ❁

Textual Memory of Right Brain Mimesis Bioculture

> Of men who came into the world, those who were cowards or led unrighteous lives may with reason be supposed to have changed into the nature of women in the second generation . . . The race of birds was created out of innocent light minded men who, although their minds were directed toward heaven, imagined, in their simplicity, that the clearest demonstration of the things above was to be obtained by sight . . . The race of wild pedestrian animals, again, came from those who had no philosophy in any of their thoughts never considered at all about

> the nature of the heavens, because they had ceased to use the courses of the head, but followed the guidance of those parts of the soul which are in the breast . . . The fourth class were the inhabitants of water; those were made of the most entirely senseless and ignorant of all . . . because they possessed a soul which was made impure by all sorts of transgression . . . Finally, trees, seeds and plants, partake of the third kind of soul, which is said to be seated between the midriff and the navel, having no part in opinion or reason or mind, but only the feelings of pleasure and pain and the desires which accompany them. For this nature is always in a passive state, and is not endowed by nature with the power of revolving in and about itself, repelling motion from without and using its own, in such a way as to observe and reflect upon any of its own concerns. Wherefore it lives and does not differ form a living being, but is fixed and rooted in the same spot, having no power of self motion . . . These are the laws by which all animals pass into one another, now as in the beginning changing as they lose or gain wisdom and folly. (Plato *Timeaus*)

The Left Brain Mimesis Bioculture and the Afterlife

With the shift to Left Brain Mimesis, biocultures begin to posit such conceptual realities as "eternal life"of both individual body and soul, in some afterlife place where all resurrected body/souls go, upon death. Of course, if all souls go somewhere other than here in new form, a hierarchy must again be established to create immortal spaces to house these immortal forms. Thus, the birth of Hell, Purgatory, Heaven and the like. Bioculturally speaking, then, individuals will upon death share the same dimensional reality and vibrational frequency that they did here, only all people will not be all together. There will be some sort of an immortal cut off point where all like will be with the like, forever, for eternity. Presumably then, the end of the world would yield a tripartate separation of mortals plus the space of God: each of the three lower brains, plus the all knowing left brain, perhaps?

❁ ❁ ❁

Textual Memory of Left Brain Mimesis Bioculture

Anyone who breaks one of the least of these commandments and teaches others to do the same will be called least in the kingdom of heaven, but whoever practices and teaches these commands will be called great in the kingdom of heaven. For I tell you that unless you righteousness and the teachers of the law, you will certainly not enter the kingdom of heaven (Matthew 5:19-20)

The Logos Bioculture and the Afterlife

This will be a short discussion, because there is no afterlife in the model of the Logos bioculture. When the machine breaks down, it's broken for good. End of story. So, a logos person will do everything to preserve the machine he/she has at all costs. The closest one can come to immortality is to create some object of technology or art or literature that will live on even though they certainly will not. The end.

❁ ❁ ❁

Textual Memory of the Logos Bioculture

This book, but only for me. Do you want to remain "immortal" after you die? Write your own book.

❁ ❁ ❁

Conclusion: Biocultural Life and Death

The Maia bioculture is formulated on a goddess whose unity is based on her triple aspects of creation, protection, destruction. Likewise, individuals model their afterlife on the underlying unity of these motions constantly interchanging with one another, now appearing as invariant forms, now dissolving back to the one. The human life then, in replication of this model, is seen as an ever present transformation through the invariants of only action and inaction, with short interludes of disassociated association with apparent form, that other biocultures call their lives.

The Mythos bioculture is born of the "harmonia" of the gods, now present, now absent, each as individual tones on the tuning board of the cosmos. Likewise, are individuals a microcosmic harmonia of these gods, mirroring the instrumentation of the cosmos in their own human matrices. When there will be no human to worship, the gods will cease to be. So, remember, remember, says the mythos man, and live in harmonia so that all the gods shall also live.

The Right Brain Mimesis bioculture creates a cosmic demiurge to decide upon the combinations of forms in the universe, though he, of course is devoid of this fate of transformation as he already permeates all form. Likewise the right brain mimctic human maintains his own integrity of self while submitting to various transformations in the course of earthly existence. Finally though, all forms shall return to their highest manifestation: Pure light; that is, the light with which we see: God.

The Left Brain Mimesis bioculture also recognizes a God who directs the course of human affairs, but this God has no affinity with the rest of creation except that, through his words, he gave it life. Likewise, humans of this bioculture direct their behavior by following the words set down in an effort to affirm or deny the direct will of God: Thou shalt, thou shalt not. This is the god of the biological left brain whose function is to allow the other brains freedom to create images. In the end, this God of yes/no will rule himself alone, as the end of the world will yield a tripartate hierarchy in such a way that no more creation can take place because none of the vibrational patterns of the different brains will be allowed communication in the "afterlife."

Finally, the logos bioculture removes these lower brains' input entirely, and the God of Left Brain Mimesis now becomes the slave to an external god made by itself – a sort of return to a Promethean/Frankensteinian world where the God of the left brain, by his persistent aggression, yields a technological revolt and, in the process, kills even this god, and with it, human's final power of creation -- thinking. The God of thought will die and with his death, the Logos bioculture shall fall back to the eternal abyss of NO-THING.

✿ ✿ ✿

APPENDIX 2.2

Membering and Dismembering: Mythopoetic Creation Technologies

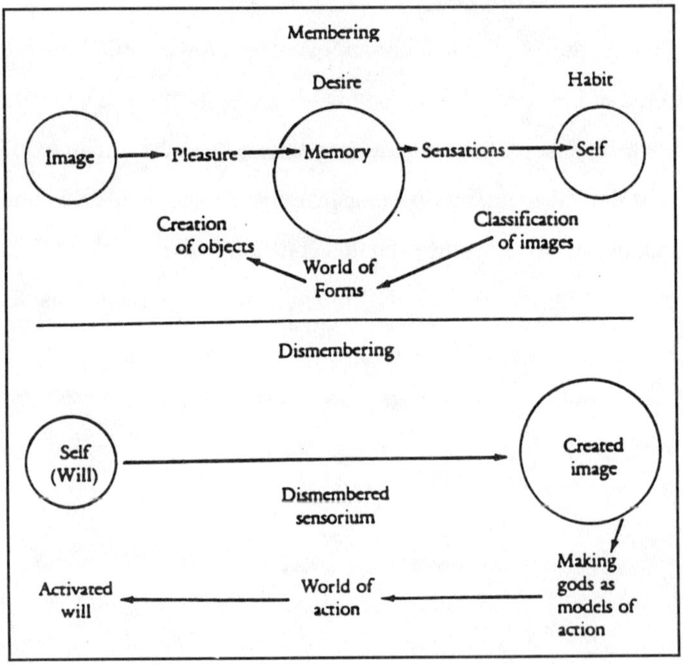

APPENDIX 2.3

Schematic Diagram of Rg Vedic Creation (All Within ASAT)

THE ONE	(Tvastr-Savitr-Visvarupa)	existed in watery
	(former-impeller-all form to be)	ASAT (Chaos)

MOVEMENT caused by the One breathing creates

TAPAS (Desire) arising in the One, Then

The One EXTENDS DOWN as the axis of creation as

AJA EKAPAD (unmanifest one-foot)

This extension causes direction and patterning. The pattern is the PILLAR.

At the AXIS of the Pillar arises Visvarupa which contains within itself Tvastr and Savitr. On the TRAJECTORY LINE OF THE ASAT AND THE ONE, Visvarupa BECOMES, in the form of a sphere called "BULLCOW". This will become the source of dualities such as bull and cow, sky and earth, BUT NOW THERE ARE NO OPPOSITES.

Next, the "matrix" for polarity is formed, with the following TWO perspectives regarding ALL polarities identified:

Perspective One: the polarity created as each opposite relates to the other (ex. bull TO cow; sky TO earth)

Perspective Two: the polarity created as each relates to the unity before the image possibility as duality is perceived (ex. bull To bullcow; cow To bullcow; Sky to earthsky; Earth to earthsky) This PERSPECTIVE IS THE "VISION" OF THE SEERS.

NOTE: These two perspectives in fact relate to the two ways that our neocortex apprehends information; perspective One, through the left hemisphere and perspective Two, through the right hemisphere.

❁ ❁ ❁

The Four Cosmic Dualities Before Manifestation (In Asat)

The first duality is MOVEMENT: Myth of two children of Tvastr

(perspective two)	(perspective one)
SARANYU (running waters) and her TWIN	VISVARUPA (bullcow)
wants to reveal her nature	wants to remain stagnant and hold all powers within himself (the sphere)
mates with VIVASVAT (insight, illumination, light)	The hero/god Indra is the one who breaks him open, releasing the (cow's) powers so that FORM could be. Indra ONLY releases the powers ("female") to make MERE FORMS.

Together, creation is possible.

They have:

Twins: YAMA/YAMI

ASVINS and (with a matrix of Saranyu)

MANU, the ancestor of humans

❁ ❁ ❁

The Four Cosmic Dualities Before Manifestation (In Asat)

The Second Duality is what holds MOVEMENT IN A MATRIX: Yamayami

(perspective two)	(perspective one)
YAMAYAMI (immortality) and her TWIN	YAMA (death/love)
Is desire which retains immortal desire. It is a hermaphroditic matrix of immortal desire which wishes to remain as a unity.	Chooses to split from unity, at times. Is the generative force and the fire of love and also of death because he chose to separate from the unity of immortality. He does this because he wants there to be self consciousness of the gods.

By the act of separation, YAMA creates a new MATRIX form, mortality; by leaving his "body" as the path though which is immortal, immortality can be "experienced". In other words, the Matrix he made was mortal; the PATH back and forth is IMMORTAL.

❁ ❁ ❁

The Four Cosmic Dualities Before Manifestation (In ASAT)

The Third duality is what HOLDS MATRICES IN MOVEMENT: ASVINS

The Asvins are twins who drive a two wheeled chariot.

(perspective two)	(perspective one)
ONE WHEEL is motionless, because it transcends even motion	ONE WHEEL circumnavigates the parameters of the possibilities of mvt., thereby creating SPACE which is still

inside NO SPACE (ASAT.)

The tension creates the duality of the stillness of the ALL and also the PARAMETERS OF MOVEMENT within the ALL. They also hold DIRECTION, BEFORE POLARITY. Saranyu gave birth to them AFTER the gods made her matrix but BEFORE her matrix gave form to MANU. Thus, the SPACE of the Asvins is "before and after matrices are made" AND THE PATH of matrices in movement.

❁ ❁ ❁

The Four Cosmic Dualities Before Creation (In ASAT)

THE FOURTH DUALITY IS FROM MATRIX TO IMAGE: MANU (dual capacity)

When the gods made an image of Saranyu on the matrix made by Yama, she created her own duality:

Saranyu (unform) and Saranyu (in form)

MANU is born of VIVASVAT and SARANYU (in form)

MANU therefore mediates in a double capacity as his own duality:

(perspective two)	(perspective one)
MANU IN FORM MATRIX	MANU AS IMAGE
creativity/body unity	creativity and body separated when
united after death	alive

BETWEEN THESE POLARITIES LIE:

PUSAN:	son of Asvins/ source of mortal mvt. in Asat
RHBUS:	Children of SAVITR/source of immortal/mortal duration AND the path between the two for humans.

MANU AS IMAGE: Is conditioned by the realm of VISVARUPA

MANU AS MATRIX: Is conditioned by the path of either:

Perspective Two: IMMORTALITY (if Yamayami is the focus)

Perspective One: LOVE/DEATH/LOVE CYCLE (if path of Yama is focus)

APPENDIX 2.4

Textual Samples of Creation Myths

Note: an asterisk (*) denotes where cosmos appears in the text.

Sample One: Rg Veda Hymn of Creation (10.129)
(translation A. de Nicolas)

Bioculture: **MAIA**

Hymn of Creation

Then, there was neither existence nor non-existence,
Neither space nor the sky spreading beyond;
What moved? Where? In whose protection?
Was there water, unfathomable, deep?
Then, there was neither death nor immortality,
Nor any sign of night or day.
Breath was then, without air, by self impulse of the One;
Other than that, there was nothing beyond.

The darkness was covered by the darkness
With no separating sign, all was water in chaos;
That One that had been covered by emptiness
Through the heat of desire (tapas) came to life.

In the beginning there was desire in that One,
That was the first seed of the mind;
The poets searching their hearts with wisdom
Found the link of existence in non-existence.

Their dividing line extended across. *
Was there a below? Was there an above?
There were those planting seeds, there were powers.
There was force from beneath, thrust from above.

Who really knows? Who can here say when was it born
Or from where creation came?
The gods come later with the creation of the universe.
Therefore, who knows from where it came into existence?

From where did creation come?
Perhaps it made itself, perhaps not,
The one looking at it from above Alone knows, unless . . . not.

Sample Two: The Creation from the Theogony of Hesiod
(translation, R. Hathorn)

Bioculture: MYTHOS

Creation

First was chaos and afterwards there were broad-breasted Gaea(earth), the stable abode of all things everlasting, and misty Tartarus, and limb-loosing Eros (love), most beautiful among the deathless gods. From chaos were born Erebus and Black Night, who mingling in love with Erebus,* conceived and bore Aether and Day. And Gaea first bore starry Uranus (sky) to be equal to herself and to cover her everywhere, so that there might be a stable abode everlasting for the blessed gods . . . And she mated with Ouranos and deep-eddying Oceanus and the Titans, the youngest and fiercest of these Titan offspring being crooked-thinking Kronos . . . But Ouranos hated his children and pushed them back into the recess and womb of Gaea . . . Gaea groaned within herself, and she devised a scheme. Creating grey flint, she fashioned a large sickle and . . . sent forth (Kronos) . . . who, holding in his hand the long and mighty jag-toothed curving blade, at one blow reaped away the genitals of his father and threw them behind him . . . Then (Kronos) became lord over his fellow Titans . . .

Sample Three: Ovid's Metamorphoses
(translation M. Colavito)

Bioculture: RIGHT BRAIN MIMETIC

Creation

Before sea, and land and sky overhead,
All nature was one appearance enclosed
In a circle, they call it chaos,
A rough and disordered mass, containing,
Nothing but stagnant weight whose seeds
Inharmoniously together were heaped.
No Titan supplied to this world brilliant light,
Not the new waxing horn of Phoebe, the moon,
Nor was the earth set into balance,
Weighted within the surrounding air,
Nor had ocean stretched out her long arms,
Enclosing within her the bordering land.
And although the sea, and land and air were there,
The land could not support its weight,
The waters lacked fluidity,
And sky was left in need of light.
Nothing was able to maintain its form
Because each thing was at odds with the other.
Contained in this sphere, heat fought with coldness,
And moisture with dry, and hard fought with softness

And things having weight fought with things having none.
And then some god, a higher nature, *
Stopped this controversy and cut,
The sky from land and from land the water,
And the thickened air was distinguished from clear.
Then after he had detached that heap,
And divided it up in its proper order,
He bound each part tight into place,
Forming a peaceful harmony.
The fiery weightless force of the heavens
Had sprung itself into the uppermost region,
And next there was air, floating in that space,
And earth, with its heavier parts fell below.
Then flowing water possessed the last place,
Embracing the earth, holding in land,
And in this way that god who arranged it,
After cutting it up, returned it to order.

Sample Four: the Egyptian Creation of the Gods and the World
(translation W. Budge)

Bioculture: LEFT BRAIN MIMESIS

The Creation of the Gods and the World

The Book of knowing the evolution of Ra, overthrowing Apep. The words of Ra which he spoke after he had come into being: "I am he who came into being in the form of Khepera; I was the creator of all that came into being . . . After my coming into being, many were the things that came into being, coming forth from out of my mouth.

For heaven did not exist; earth had not been made, nor the things of the earth . . . I raised them up out of Nu (the primeval abyss) from the state of inactivity.

I could not stand upright, so I worked a charm in my heart. I laid a foundation in Maa and made every attribute. I was alone, for I had not yet spit in the form of Shu, I had not yet emitted Tefnut, I had no partner . . . I had union with my clenched* hand, I joined myself in embrace with my own shadow, I poured seed into my own mouth, I sent forth issue in the form of Shu, I sent forth moisture in the form of Tefnut . . . They make my "eye" weak because three times they proceeded from me and from one god I made three with Shu and Tefnut and this earth came into being."

APPENDIX 3.1

The Origin of Greek Thebes

Zeus = Io (became Egyptian goddess, Isis)

I

Memphis = Epaphus (founded Egyptian Memphis)

I____________________________________I

Poseidon = Libya (founded ancient Libya) Lysianassa = Poseidon

I______I I

Belus Telephas = Agenor (king of Tyre) Busiris (king of Egypt)

I

I____________________I

Europa: carried to Crete; founder of Minoan Civilization
Thasus: settled on the island Thasus in N. Aegean Sea
Cilix: founded Cilicia, present Turkey
Phoenix: founded Phoenicians on coast of N. Palestine
Cadmus: founded Greek Thebes

APPENDIX 3.2

The House of Cadmus

Cadmus = Harmonia (Child of Ares/Aphrodite)
I
(Four Daughters/One Son)

Autonoe:	Had Actaeon (killed by his own hounds); she died
Ino:	Had Learchus/Melicertes (she killed them); became goddess, Leucothoe
Semele:	Had Dionysos; she was consumed by lightning
Agave:	Had Pentheus (she dismembered him); married to Echion, one of the Spartoi
Polydorus:	Son married daughter of Agave (Nycteis); died young, never inherited throne

```
                 Agave = Ehion
                       I
      I______________________I
Antiope = Zeus                Nycteis = Polydorus
  I_____I                          I
Zethus    Amphion = Niobe         Labdacus (died young)
                    I              I
                                      One of Spartoi
    Many Children        Laius = Jocasta_______I______Creon = Euridice
    (All but 2 killed)          I =
                             Oedipus
```

Children of Creon/Euridice:

Haemon:	Killed by Sphinx
Menoceus:	Leaped from walls of Thebes as sacrifice to Ares
Haemon 2:	Killed himself

Children of Oedipus/Jocasta:

Antigone:	Killed herself
Eteocles:	Killed by his brother
Polyneices:	Killed by his brother
Ismene:	Dies without heirs

Fate of Others:

Creon:	Murdered
Euridice:	Killed herself
Jocasta:	Hanged herself
Laius:	Murdered by son, Oedipus
Oedipus:	Blinded himself; eventually ascended into Heavens

APPENDIX 3.3

Comparative Alphabetic Systems

LATIN LETTER	A	B	M	N
PROTO-SINAITIC PICTOGRAM, 18th c. B.C.E.				
CANAANITE MEANING	"ALPU" OX	"BETU" HOUSE	"MAYYUMU" WATER	"NAHASU" SNAKE
CANAANITE, 13th c. B.C.E.				
UGARITIC CUNEIFORM, 14th c. B.C.E.				
PHOENICIAN, 13-11th c. B.C.E.				
EGYPTIAN				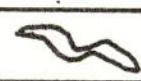
EARLY GREEK				
ETRUSCAN			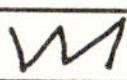	
OSCAN, 3rd c. B.C.E.				
EARLY ARAMAIC				
HEBREW OSTRACA, 6th c. B.C.E.				

APPENDIX 4.1

Who's Who of the Mind/Mind Split:
Some Key Figures in History

Akhenaten (Egyptian Pharoah; the "historical" Oedipus)

-For changing the powers of the sun into the physical object of the sun in the sky (aten), and then defacing all of the older temples, removing all the other images of the gods and replacing them with his solar symbol:

St. Augustine (Church father from Phoenician Carthage)

-For reducing the image of the Trinity to the natural operations of the mind, memory, understanding and will (*Trinitate* XIV, *Confessions* IX, X and In *De Musica*, VI)

-Also for reducing God to the size of his flesh, in this cannibalistic quote:

> *Confessions* VII,X,XVI
>
> *Cibus sum grandium: cresce et manducabis me. Nec tu me in te mutabis sicut cibum carnis tuae, sed tu mutaberis in me.*
>
> (God speaking): I am supreme food: grow up (grow into being) and you will eat me. But you will not transform me into you as you do with the food of your own flesh, but you will be changed

into me.

Gospel Writers

-For taking the Phoenician story about how Kronos (El), killed his only begotten son, Ieoud, in order to save his people from the dangers of war, and turning it into an historical account about the Christian Savior.

> Now Kronos . . . had only one son . . . and therefore they called him Ieoud. (Hebrew, "only," Phoenician, "beloved"). Even now among the Phoenicians the only son is given this name. When war's gravest dangers gripped the land, Kronos dressed his son in royal attire, prepared an altar and sacrificed him. (*Phoenician History*, PE 1.10.44=4.16.11)
>
> And the soldiers platted a crown of thorns, and put it on his head, and they put him in a PURPLE robe. And said, Hail! King of the Jews. (John 19:2) And Pilate said to unto them, Shall I crucify your king? The chief priests answered, We have no king but Caesar. (John 19:15)
>
> For these things were done that the Scripture should be fulfilled. (John 19:36)

Fulfilled, indeed.

Abraham (Biblical Patriarch)

-For instituting the "testament" of swearing under the "thigh," not to mention HIS intention to follow "the Lord" and sacrifice his son, Isaac.

> Put thy hand under my thigh, and I will make thee swear by the Lord . . . (Genesis 24:2-3)

Moses (Biblical Patriarch who shares Oedipus' life story)

-For instituting circumcision amongst the Hebrew people.

And Moses took his wife and his sons, and set them upon an ass, and he returned to the land of Egypt: and Moses took the rod of God in his hand . . . And the Lord said thou shalt say unto Pharoah, Thus saith the Lord, Israel IS my son, EVEN my first born: And I say unto thee, let my son go, that he may serve me: and if thou refuse to let him go, behold, I will slay thy son, even thy firstborn. And it came to pass by the way in the inn, that the Lord met him, and sought to kill him. Then Zipporah took a sharp stone, and cut off the foreskin of her son, and cast IT at his feet and said, Surely a bloody husband art thou to me. So he let him go; then she said, A bloody husband thou art, because of the circumcision. (Exodus 4.20-29)

Columbus (Explorer)

-For using the Phoenician maps and trade winds to discover (again) the New World in order to exploit it.

Sir Walter Raleigh, Francis Bacon and the School of Night

-For re-introducing the Phoenician "ideals" of imperialism into England. And of course, for the New Atlantis Experiment called, America.

Sigmund Freud (father of psychoanalysis)

-For universalizing the mind/mind split as the sole criterion for intelligent, psychologically healthy individuals. And of course, for the Oedipus Complex which we all don't share.

Arthur Evans (archaeologist)

-For terming "Minoan" the clearly Caananite culture of Crete, because of a myth he once heard. Also for failing to see the overwhelming similarity between Linear A and Ugaritic Script.

(PLACE YOUR OWN CANDIDATES HERE, AS YOUR RESEARCH DEVELOPS.)

NOTES

Introduction

1. I mean specifically of Paul MacLean's Triune brain; Sperry & Gazzaniga's work on split brain patients, and particularly the role of the "interpreter module" of the left brain, as outlined by Gazzaniga. See bibliography for specific references.

2. Following MacLean's model of the triune brain, the reptilian brain, also known as the R-complex, monitors involuntary activities in the body such as breathing. See appendix for detailed chart. See also, Pearce, J.C., *Evolution's End*, (San Francisco: Harper, 1992, Chapter Five) for a readable version of this model for the non-scientist.

3. Located inside the temporal lobe and part of the mammalian brain, the limbic system is the seat of memory, emotion and smell.

4. Surrounding the mammalian brain, the neo-cortex, or cerebral cortex, is divided into two hemispheres, right and left. This portion of the brain balances the influences of the reptilian and mammalian systems. Each hemisphere is believed to be associated with different human functions.

5. The right hemisphere of the neocortex is associated with nonverbal visual patterning and musical ability and emotion. It is the older of the two Hemispheres and thus has more direct contact with the two lower brains, serving somewhat as the regulator brain between them.

6. Specifically, the left hemisphere regulates verbal skills, abstract reasoning and number comprehension. It functions in a linear sequential mode, unlike the "gestalt" mode of the right hemisphere.

7. The left hemisphere, aside from being the newest, is also curiously, the largest.

8. Hesid's myth of the Races of Man is the source I am referring to here.

9. By language here I mean simply the communication between or among individuals; including, but not limited to: body language, images, sign language, verbal and written (both semantic and pictographic modes) as well as symbolic (including numerical systems, and the like), and abstract (without the medium of images, conceptual.)

❁ ❁ ❁

Chapter One

1. I mean here that most were left handed and that, based on the results of the HMI (Hemispheric Mode Indicator) test, the class as a whole, favored right brained propensities.

2. For example, the works of Walter Ong, Clancey, E. Havelock and others. The language of images I refer to here has been elaborated upon in detail by Antonio T. de Nicolas (see Bibliography for particular works.)

3. Mythopoetic is a term I use to separate a mindset of people, specifically those whose primary mode of language is imagistic. That is not to say that in the COMMUNICATION of the language they do not use the representational systems of ideographic language; simply, that the ORIGIN of their conscious experience is first, the image, and only later, the symbol.

4. Ideographic refers to the internal language system that some individuals espouse, comprising of words and concepts as the primary mode of communication, without conscious reference to the imagistic language which precedes it.

5. I call it "biocultural" because I am asserting that cultures are biologically conditioned by the parameters of the neurological functionings, and only distinguishes themselves, one from the others, by the excessive use of one function, or a combination of functions, over the others. By repetition of these dominant functions, human diversity (and within the culture, homogeneity,) occur, thus MANIFESTING ITSELF AS CULTURE. See Luria, A.R., *The Neuropsychology of Memory*, (N.Y.:Halstead/Wiley BMB Themepack

#9,1977)

6. This is the epistemological origin of the doctrines of reincarnation and metempsychosis, expounded upon by such western philosophers as Pythagoras and Plato.

7. See, Gazzaniga, M., *Nature's Mind*, (N.Y.: Basic, 1992, Chapter Five.)

8. Don't think of pink elephants. See what I mean?

9. Pearce, p.176

10. Since the left hemisphere doesn't need to rely on the sensory data as supplied by the other brains, it is free to conceptualize without body referent. Once it does, it creates its own substitution system (a new language, if you will) to catalog these "discoveries", thereby removing the need for the other classification systems in the brain: the language of images produced by the union of the sensory input as produced by the other brains and made conscious of by the right hemisphere.

11. From MacLean, Paul, "On the Evolution of the Three Mentalities of the Brain", in Neuman, G. ed. *Origins Human Aggression*, (N.Y.: Human Sciences Press, 1986, p.29)

12. Ibid, p.30

13. Ibid, p.33

14. Ibid, p.34

15. Ibid, p.38-40

16. Ibid, p. 37

17. Pearce, p. 76

18. MacLean, p.40

19. Pearce, p.182

20. Restack, R. *The Brain*, (N.Y. Bantam, 1984, p. 256)

21. Ibid, p.259

22. Pearce, p.176

23. Gazzaniga, pp.102-104

24. Ibid, p. 134

25. Ibid, p.125

26. Pearce, p.183

27. Gazzaniga, p.2

28. Pearce, p. 99

Chapter Two

1. This line is from the *Rg Vedic Hymn of Creation* (10.129, translation, de Nicolas). See the appendix for the complete hymn.

2. For a comprehensive analysis of the nature of mythopoetic creation myths, see, Colavito, M. *The New Theogony*, (N.Y.: SUNY Press, 1990)

3. The study of memberment and dismemberment as it relates to acts of imagining were first described by Antonio T. de Nicolas in his *Meditations Through the Rg Veda*, (N.Y. Nicolas-Hays, 1976). In his subsequent work, *Powers of Imagining*, (N.Y.: SUNY Press, 1986) he elaborates upon how this

technique is utilized for spiritual practice. See also appendix, for a chart on how this act of dismemberment is performed in mythopoetic biocultures.

4. Ovid, *Metamorphoses, Book One*, (translation, Colavito)

5. R.V. 1.164:38, (translation, de Nicolas)

6. *Hesiod, Works and Days*, 47-105 contains the entire story.

7. The popular notion of sacrifice as "surrendering" is not what is meant here. The original meaning of the word is to "make Holy"; thus, to offer up to God/the gods and thereby sanctify the victim of sacrifice, be it an animal or human. (ex. the god-kings of the Near East who were deified "made holy" by sacrifice.) Our meaning of the word has come to be synonymous with the left brain mimetic meaning because of our cultures' Judeo-Christian slant, which is a product of a left brain mimetic bioculture, as you shall see in the rest of this chapter.

8. I am analyzing Genesis 3.

9. In Bandler and Grinder, *Trance-formations* (Utah: Real People Press, 1981), the following quote summarizes how it is the left brain that creates negation because ,"negation only exists in language and not in experience"(p.67). For example, to say, don't look at this page, the mind must first create the experience of looking and then negate it. This can only be performed by a conceptual exercise, which we know has its origin only in the left hemisphere.

10. Every religion has its sects of asceticism for those "holy" individuals for whom self sacrifice means, "beat up my (evil) body." The Christian martyrs and Hindu Sadhus are two examples.

11. Hawking, S., *A Brief History of Time*, (N.Y.: Bantam, 1988, p.11)

12. See, de Nicolas, A. *The Biology of Religion: The Neural Connection Between Science and Mysticism*, (Tokyo: International Buddhist Study Center,

1990) for a biological examination of the roots of this Self, and its implications.

❁ ❁ ❁

Chapter Three

1. The answer is BEE LEAF (belief!)

2. The chart I am referring to at the beginning of the chapter is artificial in that it doesn't represent any particular culture's actual development; however, by theoretical analysis, the MOVES are correct and consistent with what we know of the development of language systems over time. See, Hooker, J.T.,et al, *Reading the Past,* (Berkeley: University of California Press, 1990).

3. I contend that this did not occur as a result of some "patriarchal invaders" coming in from the outside to take over the matriarchal societies. My contention is that IDEOGRAPHY ITSELF was the patriarchal invader who took over from the inside through acculturation with the innocent introduction of left brain mimetic biocultures to mythopoetic ones, through trade.

4. See Walker, B., *The Woman's Encyclopedia of Myths and Secrets*,(San Francisco: Harper Collins, 1988, "Moon"; "Motherhood") for more striking evidence that even what I've quoted here.

5. Walker, pp.684-86. The most comprehensive evidence of the role of Gematria in left brain biocultures is, Fideler, D., *Jesus Christ, Sun of God,* (Illinois: Quest Books, 1993).

7. "In the beginning was the Word, and the Word was with God, and the Word WAS GOD. The same was in the beginning with God." (emphasis added.) John 1:1-2)

8. Other characters besides Oedipus and his family, who share this logos hero paradigm include Sargon of Akkadia; Moses; Samson; Mithra; and, of course, Akhenaten. It is interesting to note the blend of mythic and historical

figures in this list. Is this a clue? Yes. Read on, and you will see why.

9. The mythic traditions that follow are taken from a compendium of sources. For a list of all the classical references to the stories, see Tripp, E., *The Meridian Handbook of Classical Mythology*, (N.Y.: Meridian, 1970). For the best narratives of the stories, and the ones I most closely follow, see Hathorn, R. Greek Mythology, (Beirut: American University of Beirut Press, 1977).

10. Tyre is in Phoenicia.

11. In other words, the Minoan Civilization, according to the mythic tradition, was founded by Phoenicians.

12. See appendix for complete myth.

13. The analysis that follows regarding the interpretation of these two words is taken from Hathorn, Greek Mythology, chapter XVI.

14. See, Walker, B., *Woman's Encyclopedia*, "Castration", for all the details.

15. Curiously, the name does appear (in feminine form, Poseidaia) in Linear B tablets, from the allegedly "Mycenaean" times. (Hathorn, p.112)

16. Dating of the hymns of the Rg Veda range from 5000 B.C.E. to1500 B.C.E. I prefer to place the hymns sometime in the third millennium B.C.E., for reasons that will become obvious as the chapter progresses. Further, while this analysis of the Rg Vedic Creation myth is indeed based on a compendium of hymns, re-interpreted, two secondary sources proved invaluable in the completion of this interpretation. First, the article by Stella Kramrisch,"Two: Its Significance in the Rg Veda" (in *Indological Studies*, New Haven: American Oriental Society, 1962, 109-136) provided the overall mythic framework that I follow intact here. Also, de Nicolas, A, *Meditations Through the Rg Veda*,(N.Y.: Nicolas-Hays, 1976) provided the epistemological framework that gives credibility this interpretation. See appendix for a

Schematic diagram of this myth.

17. It is my contention that Indra represents the intrusion of the left brain bioculture in this myth system. He is completely and entirely ancillary to the Rg Vedic cosmological system and his name doesn't even have meaning in the language. In other words, no one knows where Indra came from. And eventually, his cult disappeared and no one seems to know why. My guess? He was the "Marduk/Baal" of the Rg Veda, for those traders/foreigners of the left brain biocultures who happened upon Indian shores and chose to stay for a while.

Chapter Four

1. These quotes were taken from, Harden, D., *The Phoenicians* (N.Y.: Praeger, 1962)

2. See, Attridge and Oden, *Philo of Byblos, The Phoenician History*, (Washington D.C.: The Catholic Quarterly Monograph Series 9) for a complete account of the transmission of this text.

3. Attridge and Oden, p. 33

4. Ibid, p. 57

5. Ibid, p. 55

6. Genesis 6-10

7. For this and all later references to this explosion, I am relying on the scientific reports as described in Pellegrino, C.,*Unearthing Atlantis*, (N.Y.: Random House, 1991)

8. Pellegrino, Part One

9. Pellegrino, and archaeologist Marinatos before him make the argument

that it was actually the Minoan civilization that Plato was referring to in his myth of Atlantis; however, I believe that the weight of the evidence will show that there is no such thing as the "Minoan Civilization"- that it was merely a part of the ancient Phoenician alliance.

10. Plato, Critias, 119c-120c (translation, Jowett).

11. Ibid, 115d-116c

12. See, Harden, Chapter Two, for a complete account of Phoenician geographic layouts.

13. See, Walker, B., "Kingship".

14. Quoted from, de Nicolas, *Meditations Through the Rg Veda*, (N.Y. Nicolas-Hays, 1976, p. 148)

15. This thesis has been proposed and articulated by Antonio T. de Nicolas in *Habits of Mind*, (N.Y.:Paragon House,1989)

16. See, Walker, B., "WEMS, *Judas*. The reader is advised that the author is aware of the striking nature of the discussion about Jesus and Judas that is about to follow. Please be forewarned that this discussion is suggested merely to drive home the extent to which this pervasive myth of the logos hero has infiltrated even Biblical exegesis. I am not suggesting THAT this interpretation is fact: I am suggesting that this interpretation IS FACT for those followers of Christianity who are of the Logos Biocultural type. For more about the specific difficulties in the nature of the character of Judas himself, see, Baigent, M., *The Messianic Legacy*, (N.Y.:Dell, 1986). For more information on the myth of Judas as described in the text, see Rank, O., *The Myth of the Birth of the Hero*, (N.Y.:Vintage Books, 1959).

17. See, Baigent, M., *Holy Blood, Holy Grail*, (N.Y.:Dell Publishing, 1982, 365-369)

BIBLIOGRAPHY

Attridge Harold and Oden, Robert. "Philo of Byblos: The Phoenician History." Washington D.C.: *The Catholic Quarterly Monograph* Series 9, 1981

Baigent, M. *Holy Blood, Holy Grail.* New York, NY: Dell Publishing, 1982

Bandler, R and Grinder, J. *Trance-Formations*. Moab: Real People Press. 1981

Best J. and Woudhuizen, F. *Ancient Scripts From Crete and Cyprus*. Leiden: Brill, 1988

Booth, P. and Colavito, M. *The Phaistos Disk: Written Word of an Oral Culture*. Smithtown: Exposition Press, 1982

Burkert, Walter. *Structure and History in Greek Mythology and Ritual.* Berkeley: University of California Press, 1979

----------. *Greek Religion*. Massachusetts: Harvard UP, 1985

----------. *Ancient Mystery Cults*. Massachusetts, Harvard UP, 1987

Campbell, Joseph.,ed. *The Mysteries*. Princeton: Princeton UP, 1978

----------. *The Masks of God*. Vols I-IV. New York: Penguin, 1982 ed.

----------. *The Mythic Image*. Princeton: Princeton UP, 1974 ed.

Ceram, C.W. *The Secret of the Hittites*. New York: Dorset, 1955

Chadwick, John. *The Decipherment of Linear B.* London: Cambridge UP, 1976 ed.

Childe, V.Gordon. *The Aryans.* New York: Dorset, 1987 ed.

Colavito, M. *The New Theogony.* Albany: SUNY Press, 1990

----------. *The Pythagorean Intertext in Ovid's Metamorphoses.* Lewiston, N.Y.: The Edwin Mellen Press. 1989.

Cornford, Francis. *Plato's Cosmology: The Timaeus of Plato Translated with a Running Commentary.* London: Routledge and Kegan Paul, 1966 ed.

Cotterel, Arthur. *The Minoan World.* London: Michael Joseph, 1979

Cox, Sir George. *The Mythology of the Aryan Nations.* New York: Paragon Book Gallery, 1963

Cumont, Franz. *Afterlife in Roman Paganism.* New Haven: Yale UP, 1959 ed.

----------. *Oriental Religions in Roman Paganism.* New York: Dover Publications, 1956 ed.

de Lubicz, R.A. Schwaller. *Sacred Science.* Trans Vanden-Broeck. New York: Inner Traditions, 1982 ed.

----------. *The Temple in Man.* Trans Lawlor. Vermont: Inner Traditions International, 1977

de Nicolas, Antonio. *Meditations Through the Rg Veda.* New York: Nicolas-Hays, 1976

----------. *Powers of Imagining.* Albany: SUNY Press, 1980

----------. *Habits of Mind.* New York: Paragon House, 1989

----------. *The Biology of Religion*. Tokyo; international Buddhist Study Center, 1990

de Rola, Stanislas Klossowski. *Alchemy: The Secret Art*. London: Thames and Hudson, 1973

Dicks, D.R. *Early Greek Astronomy to Aristotle. Ithaca*: Cornell UP, 1985

Dijksterhuis, E. J. *The Mechanization of the World Picture*. Princeton: Princeton UP, 1986

Doumas, C. *Thera*. London: Thames and Hudson, 1983

Fideler, David. *Jesus Christ Sun of God*. Wheaton: Quest Books, 1993

Fontenrose, Joseph. *Python*. Berkley: University of California Press, 1980

Frymer-Kensky, Tikva. *In the Wake of the Goddess*.New York: Fawcett, 1992

Gazzaniga, M. *Nature's Mind*. New York: Basic Books, 1992

Godwin, Joscelyn. *Mystery Religions in the Ancient World*. San Francisco: Harper and Row , 1981

----------. *Harmonies of Heaven and Earth*. Vermont: Inner Traditions International, 1987

Gonda, J. *Notes on Names and the Name of God in Ancient India*. London:North Holland Publishing Co.,1970

Gordon, Cyrus. *Forgotten Scripts*. N.Y.: Basic Books, 1982 ed.

----------. *Ugarit and Minoan Crete*. New York: Norton.1966

Green, Miranda. *The Sun-Gods of Ancient Europe*. London: Hippocrene, 1991

Guthrie, W. K. C. *A History of Greek Philosophy: 1*. Cambridge,1962

----------. *The Greeks and Their Gods*. Boston: Beacon Press,1967 ed.

Hamilton, Edith and Cairns, Huntington, eds. *Plato: The Collected Dialogues.* Princeton: Princeton UP, 1982 ed.

Harden, Donald. *The Phoenicians*. New York: Praeger, 1962

Harrison, Jane Ellen. *Themis.* New York: University Books, 1966 ed.

Hawking, Stephen. *A Brief History of Time.* New York: Bantam, 1988

Herm, Gerhard. *Los Fenicios. Barcelona: Ediciones Destino*, 1973 ed.

Hooker, J.T. et al. *Reading the Past*. Berkeley: University of California Press, 1990

Ions, Veronica. *The World's Mythology*. London: Hamlyn Press, 1974

Kaiser, W. C. *The Ugaritic Pantheon*. Diss. Brandeis U. 1973

Knight, Richard and Wright, Thomas. *A History of Phallic Worship.* New York: Dorset, 1992 ed.

Kramrisch, Stella. "Two: Its Significance in the Rg Veda." *Indological Studies.* New Haven: American Oriental Society, 1962

Lawlor, Robert. *Sacred Geometry.*New York: Crossroad, 1982

Levy, Ernst. *A Theory of Harmony.* Albany: SUNY Press, 1985

-----------. and Levarie, S. Tone: *A Study in Musical Acoustics*. Ohio: Kent State UP, 1980

Mallory, J. P. *In Search of the Indo-Europeans.* London: Thames and Hudson, 1989

Mookerjee, Ajit. *Kali: The Divine Feminine.* New York: Destiny Books, 1988

McClain, Ernest. *The Myth of Invariance.* New York: Nicolas Hays, 1976

-----------. *The Pythagorean Plato.* New York, Nicolas Hays, 1978

Moscati, Sabatino, ed. *i Fenici.* Milano: Bompiani, 1988

Muller-Ortega, Paul. *The Triadic Heart of Siva.* Albany: SUNY Press, 1989

Neuman, G. ed. *Origins of Human Aggression.* New York: Human Sciences Press, 1987

O'Meara, Dominic, ed. *Neoplatonism and Christian Thought.* Albany: SUNY Press , 1982

Ovid. *Metamorphoses.* Trans. Mary Innes. Baltimore: Penguin Books, 1955

Pearce, Joseph Chilton. *Evolution's End.* San Francisco: Harper Collins, 1992

Pellegrino, C. *Unearthing Atlantis.* New York: Random House, 1992

Pinel, John. *Biopsychology.* Needham Heights: Allyn & Bacon, 1990 ed.

Pritchard, James. *The Ancient Near East*. Princeton: Princeton UP, 1973 ed.

Rank, Otto. *The Myth of the Birth of the Hero.* N.Y.: Vintage Books, 1959

Restak, Richard. *The Brain.* New York: Bantam, 1984

Rohde, E. *Psyche: The Cult of Souls and Belief in Immortality among the Greeks*. London, 1925

Rose, H. J. *Religion in Greece and Rome.* New York: Harper and Row, 1959

Sargent, Thelma, ed. *The Homeric Hymns*. New York: Norton, 1975 ed.

Saunders, Jason, ed. *Greek and Roman Philosophy after Aristotle*. New York: The Free Press, 1966

Scott, Walter. Trans. and ed. *Hermetica*. Boston: Shambhala, 1985

Sebeok, Thomas, ed. *Myth a Symposium.* Bloomington: Indiana UP, 1965 ed.

Shepard, Gordon. *Neurobiology*. N.Y.: Oxford UP, 1994 ed.

Taylor, A. E. *A Commentary on Plato's Timaeus.* Oxford, 1928

Tejera, Victorino. *Modes of Greek Thought.* New York: Appleton, 1971

----------. *Plato's Dialogues One by One.* New York: Irvington, 1984

Vanderbroeck, Andre. *Philosophical Geometry.* Vermont: Inner Traditions International, 1987 ed.

Walker, Barbara. *The Woman's Encyclopedia of Myths and Secrets.* San Francisco: Harper Collins, 1983

West, M. L. *The Orphic Poems*. Oxford: Clarendon Press, 1984 ed.

Wilbur, J. B. and Allen, H. J. *The Worlds of the Early Greek Philosophers.* New York: Prometheus Books, 1979

Zeller, Eduard. *Outlines of the History of Greek Philosophy.* New York:Meridan, 1969 ed.

INDEX

M

N

O

P

R

S